The Observer's Sky Atlas

Second Edition

Springer

New York
Berlin
Heidelberg
Barcelona
Hong Kong
London
Milan
Paris
Singapore
Tokyo

E. Karkoschka

The Observer's Sky Atlas

With 50 Star Charts Covering the Entire Sky

Second Edition

Springer

E. Karkoschka
Lunar and Planetary Lab
Space Sciences Building
University of Arizona
Tucson, AZ 85721
USA

Cover photograph: Based on a photograph by the author, it shows the southwest part of the constellation Sagittarius together with the brightest Milky Way clouds. North is to the upper left. The red nebula near the left edge is the Lagoon Nebula M8. Many other objects can be identified with the help of chart E20.

Frontispiece: The sword of Orion, containing the Orion Nebula. Looking at it with a large telescope on a dark night gives one of the grandest views in the universe. The faint reflection nebula NGC 1973 lies half way up to the top of the photograph where the stellar group NGC 1981 can be seen.

With 50 star charts, 3 black and white photographs, and 6 line drawings.

Library of Congress Cataloging-in-Publication Data
Karkoschka, Erich.
   [Atlas für Himmelsbeobachter. English]
   The observer's sky atlas : with 50 stars charts covering the
entire sky / E. Karkoschka. — 2nd ed.
      p.   cm.
   Includes bibliography references and index.
   ISBN 0-387-98606-5 (softcover : alk. paper)
   1. Astronomy—Charts, diagrams, etc.   2. Stars—Atlases.
   I. Title.
   QB65.K3713   1998
   523.8′0222′3—dc21   98-29450

Printed on acid-free paper.

Title of the original German edition: Atlas für Himmelsbeobachter
© Franckh'sche Verlagshandlung, W. Keller & Co., KOSMOS-Verlag, Stuttgart 1988

Production managed by Lesley Poliner; manufacturing supervised by Joe Quatela.
Photocomposed pages prepared from the author's LATEX file.
Printed and bound by Edwards Brothers, Inc., Ann Arbor, MI.
Printed in the United States of America.

9 8 7 6 5 4 3 2 1

ISBN 0-387-98606-5 Springer-Verlag New York Berlin Heidelberg   SPIN 10659306

Contents

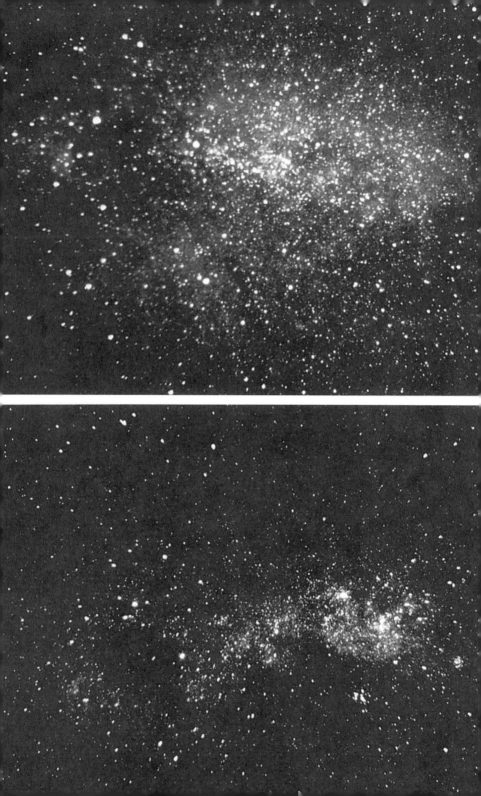

Explanatory Notes

Introduction

Can you remember being impressed by a clear starry sky? Look at the Milky Way through binoculars and it will reveal its many hundreds of thousands of stars, double stars, stellar clusters, and nebulae. If you are a new observer, it is not that easy to find your way in this swarm of stars, but this atlas tries to make it as easy as possible. So now it is not just experienced amateurs that can enjoy looking at the heavens.

Two additional observing aids are recommended. The first is a planisphere, where one can dial in the time and day in order to see which constellations are visible and where they are in the sky. The second is an astronomical yearbook listing the current positions of the planets and all important phenomena.

So, let us begin our journey around the night sky, and see what the universe can reveal to us!

Sky Atlases

Most sky atlases can be classified into one of two major groups according to the number of stars they contain. Some atlases only show the stars visible to the naked

---

Facing page, top: The constellation Cygnus (Swan) in the midst of the northern Milky Way. The photograph gives an impression of the uncountable stars in our Milky Way. This becomes more conspicuous when you sweep through Cygnus with binoculars. Under a very dark sky, one can try to find the North America Nebula, Pelican Nebula, and Veil Nebula (see p. 45). These are difficult nebulae and are only barely visible on this photograph as well. For orientation: Deneb is the bright star on the left side; Albireo is near the right edge, nearly as high as Deneb

Facing page, bottom: The region around the constellation Crux (Southern Cross) in the southern Milky Way. Aside from the Magellanic Clouds, this part is a special attraction of the southern sky. Directly to the lower left of the cross is a dark nebula, the Coalsack. It displays beautiful detail in binoculars. In the right part of the photograph is the bright Eta Carinae Nebula, surrounded by bright clusters. The star Eta Carinae illuminates the nebula and is currently not visible to the naked eye, although it was the second brightest star in the sky during two decades of the nineteenth century. It is a candidate for the next supernova explosion in our part of the Milky Way. The enormous flash of the explosion might already be on its 6000-year journey to us

Explanatory Notes

eye. As there are not more than a few thousand such stars, such charts can be simple and clear and can be arranged in a handy format. They are ideal for all naked-eye observations. The other group of atlases contains the stars visible through binoculars or telescopes. As there are a million stars within the reach of binoculars, such atlases need hundreds of charts, often arranged in several volumes. They are ideal for observations with binoculars and telescopes.

This atlas steers a middle course. It contains the whole sky visible to the unaided eye (limiting magnitude 6), and finder charts for 250 interesting objects for binoculars and small telescopes (limiting magnitude 9). Since these finder charts only cover approximately ten percent of the whole sky, it was possible to put all this information into a very convenient format.

Some atlases contain as many codes and labels as possible for each object. They are quite useful for work at home at the desk. The other extreme is represented by photographic atlases containing no labels at all. They are recommended when it comes to comparison with the real sky. This atlas again lies between the two extremes. The star charts are clear and contain just one label for important objects, since all the other data can always be found on the page facing the chart.

Catalogs

As well as a naked-eye atlas and a binocular atlas, observers also use a catalog to look up important data such as double-star separations or the magnitudes of nebulae. This atlas combines these three functions. To work with different books can be troublesome because, between them, object selection and labeling may be quite different. In this atlas all objects labeled in the charts are listed in the tables on the facing page, naturally with the same designation, and all objects in the tables are labeled in the facing star charts. This makes observing as easy as possible.

Until 1997, distances to most stars had to be estimated since there was no

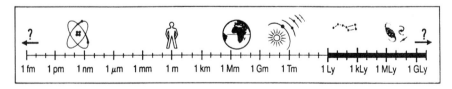

Figure 1: Between the size of an atomic nucleus and the furthest known objects in the universe, we have explored 41 orders of magnitude. This atlas contains objects further than one light-year: that means the last ten orders of magnitude. Nobody can really imagine these distances. But if we shift these ten orders of magnitude to the left, to the sizes we feel comfortable with, then we can get at least a feel for the world of stars and galaxies.

Table 1: The mean relative uncertainty of stellar distances. It has significantly improved due to the publication of the Hipparcos Catalogue in 1997. All data of this book consider this progress.

distance (light years)	10	20	40	80	150	300	600	1 200	2 500	5 000
uncertainty until 1997 (%)	3	6	12	25	35	45	50	50	50	50
uncertainty since 1997 (%)	0.25	0.5	1	2	4	8	15	25	40	50

reliable measurement. In the summer of 1997, accurate measurements of 1 000 000 stars by the European satellite Hipparcos were published. They are an important basis for the data given in this new edition of "The Observer's Sky Atlas." The reliability of stellar distances has strongly improved, especially for nearby stars as shown in Table I. Finally, we know which stars of a constellation are really close together and which are foreground or background stars. This information can be easily digested by the charts and data in this atlas.

Until 1997, many data on binaries (double stars) such as magnitude, color, and separation were based on more or less reliable estimates from observers. The Hipparcos satellite revealed previous errors and often provides data more accurate than sufficient for an observer. The new knowledge is included in this book. Other sources are: *Sky Catalogue 2000.0*, the *Yale Bright Star Catalogue*, the *Smithsonian Astrophysical Observatory* (SAO) *Star Catalog, 291 Doppelsternephemeriden für die Jahre 1975–2000, Ovschni Katalog Peremenich Zvjezd, Synopsis der Nomenklatur der Fixsterne, Délimitation Scientifique des Constellations, Burnham's Celestial Handbook*, and the *Webb Society Deep-Sky Observer's Handbook*.

Object Selection

This atlas contains 250 nonstellar objects listed under the general term "nebula": planetary nebulae, diffuse nebulae, open or galactic star clusters, globular star clusters, and galaxies. In addition to all 110 Messier objects, 140 additional nebulae that have magnitudes like those of many Messier objects have been selected. Among similar nebulae, those further north were slightly favored in the selection. All these nebulae can be observed with an amateur's telescope. Following each table of nebulae is a short description of each object for binoculars or a telescope under good sky conditions. Here the term "amateur's telescope" is considered to be a telescope with an aperture lying in the range 100–200 mm (4–8 in.). Today, many amateurs own still larger telescopes. This atlas is also useful for them, but it only satisfies part of their telescopes' capabilities.

The catalog of stars contains 900 naked-eye stars. It is complete up to magni-

Explanatory Notes

tude 4.0. There are 556 stars up to this magnitude. Most of the fainter listed stars are doubles or variables.

Many thousands of double and multiple stars are observable with amateurs' telescopes. 250 interesting ones are listed in the table of binaries. Their components are at least magnitude 8.0, their combined light brighter than 6.0. Apart from a few very close binaries, all these can be separated in an amateur's telescope.

The tables also list data for 80 variable stars visible to the naked eye. Variable stars with a variation of at least a quarter of a magnitude were considered. All those which get brighter than fifth magnitude (and a few fainter ones) are included.

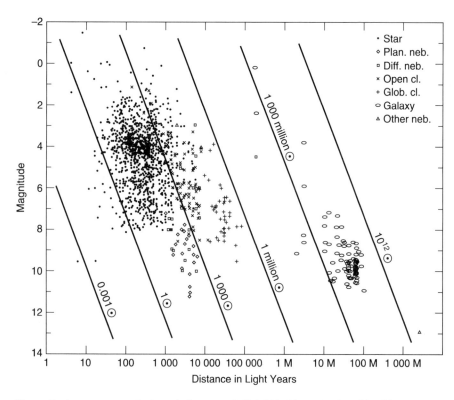

Figure 2: Apparent magnitude and distance of all 1427 objects cataloged in this atlas; binary components are indicated individually. The steep lines show the luminosity relative to the solar luminosity (if interstellar absorption is neglected).

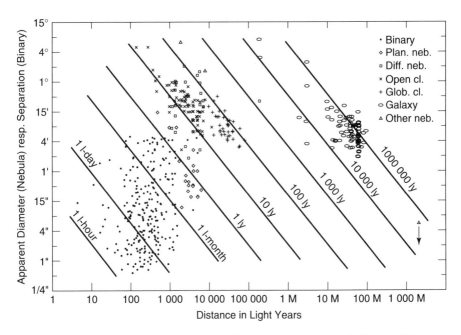

Figure 3: True size (inclined lines) as a function of apparent size and distance. The true size refers to the extent perpendicular to the line of sight. The radial extent is unknown for most objects. One light-hour is approximately 1000 million kilometers or seven astronomical units (AU).

Celestial Coordinates

In astronomy, many different coordinate systems are commonly used. To enjoy the night sky it is not necessary to tangle with the mathematics of coordinates. However, it is quite useful to become familiar with the most important coordinate system, the celestial equatorial system. One can imagine the equatorial system as a projection of the earth's longitude and latitude circles from the center of the earth onto the celestial sphere. Right ascension corresponds to geographical longitude; declination corresponds to geographical latitude. In the same way that Greenwich marks the zero meridian on earth, the first point of Aries serves as the zero point for the right ascension: it is the location of the sun on March 20/21. From there, right ascension is measured towards the east from 0° to 360°, or, more often, from 0^h to 24^h (hours) with $1^h = 15°$. Declination increases as geographical latitude from 0° at the equator to 90° at the poles. Northern declinations are positive, southern ones negative. The position of a star is uniquely determined by its right ascension and declination.

Explanatory Notes

Precession

Since the first point of Aries slowly moves across the constellations, the coordinates of the stars are constantly changing. The coordinates in this atlas refer to the standard reference frame of J2000.0. By 2015, the coordinates will have changes by 0.2° or less so that the given coordinates can be used for most practical purposes without applying any correction.

Sidereal Time

The starry sky and the celestial coordinate system complete one revolution every day. Stars at the same declination describe the same orbit on the celestial sphere. After one sidereal day of 23^h56^m, every star is back at its original position. The sidereal time indicates the rotation since the first point of Aries passed the meridian. The meridian stretches from south to north, passing through the zenith. All stars reach their highest point on the meridian. At 0^h00^m sidereal time the first point of Aries is on the meridian. At 1^h sidereal time stars with the right ascension 1^h are passing across the meridian, and so on. Knowing the current sidereal time defines the region of the sky which is visible best. Many planispheres have a dial to read off the approximate sidereal time.

Arrangement of Star Charts

In this atlas the whole sky is divided into 48 regions which are grouped into three sections: N = northern sky, E = equator and ecliptic, S = southern sky. The northern sky here means the area north of about 30° declination. From mid-northern latitudes, for example, it is clearly visible every night. The very first chart (NP = north polar region) contains stellar magnitudes to mag. 13 for estimating the limiting magnitude to the unaided eye, binoculars, and telescopes. The section for the equator and ecliptic contains declination zones where the sun, moon, and planets have their paths. Constellations in this region are only visible at certain times. Of course, they are best visible near the meridian. The sky south of −36° declination is labeled here as the southern sky. It cannot be observed north of 50° latitude. But further south more and more of the southern sky becomes visible. Northern-hemisphere observers should not miss the opportunity to observe the beauties of the southern sky when traveling south.

Within each of the three groups the charts are ordered in right ascension from 0 to 24. For example, the charts N12, E12, and S12 all display regions near 12^h right ascension. The objects in the tables are also ordered in right ascension, which increases from right to left in the charts. Furthermore, the even-numbered charts E0, E2, . . . mostly contain regions south of the equator, while E1, E3, . . . display regions mostly north of the equator. To find a particular chart quickly, consult the key chart at the end of the book.

Within each chart, bright stars and nebulae are labeled in the highlighted section. Data for these objects are listed on the page facing the chart. Objects in the grey section are labeled on other charts. Neighboring chart numbers can be found at the white-grey boundary.

In the catalog, each entry of a constellation is followed by a rectangle which represents the facing star chart greatly reduced. The object's location in the chart is indicated by a dot for the main star charts and by a small circle for the round enlargements (finder charts). This way, the object's position on the chart is located more easily than with coordinates and coordinate grids.

Directions in the Sky

On the earth we are very familiar with the direction of the four cardinal points. In the same way, directions are defined at each point in the sky. North is the direction to the celestial north pole near Polaris. West is the direction in which the sky is carried by the diurnal rotation of our planet. Therefore west is sometimes called preceding, and east is called following. When looking up at a constellation lying in the south, north is up, west is right, south is down, and east is left. We note that east and west on star charts are opposite to east and west on maps. However, in common with maps, all charts in this atlas have north at their top.

When comparing a chart with the sky it is important to know the directions in the sky in order to turn the charts until they match the sky. For comparison with rising constellations, the charts need to be turned somewhat counterclockwise, and clockwise for setting constellations. In an inverting telescope, directions are particularly important, even if they are not so clear. If you are not sure of them, just watch the motion of the stars through the eyepiece (clock drive off). They always move to the west. Further, notice that a standard diagonal gives a mirror image (if the total number of reflections is odd). You would need to look through a mirror at the charts in order to match the view in the eyepiece. Therefore the use of a standard diagonal is not recommended for deep-sky-object hunting. Diagonals with two reflections are only slightly more expensive than those with one reflection. They are the recommended choice for every work with star charts.

Size and Scale

Distance and size in the sky are measured in degrees, arc minutes, and arc seconds ($1° = 60'$, $1' = 60''$). In this atlas, declination is given in degrees, the size of nebulae

_____

in arc minutes, and the separation of double stars in arc seconds. There are no mixed entries like 8°48′ or 3′12″. This latter practice is continuing to disappear from astronomical tables.

When using star charts it is important to have an idea about the scale of the charts. The star charts in this atlas have a scale of 4°/cm (10°/in.), the round enlarged sections 1°/cm (2.5°/in.). Distances in the sky can be quite accurately estimated with your hand. If you hold it about 60 cm from your eye, 1 cm on your hand corresponds to 1° in the sky. Once you have measured some sizes on your hand, you will always have a "handy" aid present at your observation sessions.

When observing with binoculars and telescopes it is very helpful to know the diameter of the field of view. You can estimate this by comparing it with the disk of the full moon, which is about 30′ = 0.5° across. Better still are data from the manufacturer, or your own measurements. For example, a field of 5° in binoculars corresponds to a 5 cm (2 in.) circle in the round enlarged sections of this atlas. For an observer with such binoculars, a transparency with circles of 1.25 and 5 cm (0.5 and 2 in.) diameter is a helpful aid since it shows the visible field in the main and finder charts. Trying to work with star charts can be difficult for the inexperienced observer, but it becomes easy by knowing direction and scale.

All maps and charts are somewhat distorted, because the sky is spherical and charts are flat. Therefore the scale is not constant. Star charts containing a large fraction of the whole sky (e.g. p. 132) necessarily have large distortions. For all other charts the distortion is kept low by using appropriate projections. These projections show all right-ascension circles as straight lines, perpendicularly intersecting the declination circles throughout. The scale in the direction north–south (declination) is 4°/cm (10°/in.), while the scale in the direction east–west varies a little around this value as shown in Fig. 4. The round enlarged sections are stereographic projections and are practically free of distortion because of the small area of sky shown.

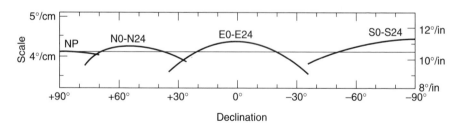

Figure 4: The scale of the star charts in the east–west direction, in the direction of right ascension (curves). The scale in the perpendicular direction (in declination) is 4°/cm everywhere (horizontal line). The difference between both scales is small; the charts are nearly undistorted.

8

Designations

There are many types of designation for astronomical objects. For the observer it is sufficient to be familiar with just the most important ones. Constellations are designated by their 88 official Latin names. Abbreviations for, and meanings of, the constellations are listed on pp. 126, 127. Bright stars are designated by the Bayer Greek letter and/or Flamsteed number, with the constellation name in the genitive, for example α Ori = Alpha Orionis = 58 Orionis. Latin letters are used for variables and stars without a Bayer or Flamsteed designation. Some stars also have names which are mostly spelled here according to their original form. Today their spelling varies in different languages. Names are not that useful for the identification of stars, except the most common ones which are printed bold in the tables. The pronunciation of these foreign names poses problems for many people who are not familiar with Latin, Greek, and Arabic. For simplicity the names are often pronounced just as if they were English. Actually, the original pronunciation is much simpler than today's English, since every letter is always used in the same way: a as [ah], e as [eh], i as [ee], u as [oo], c as [k], etc. ξ Cep is pronounced [ksee keh-feh-ee]. Of course, there is no right and wrong in pronunciation, just as a dialect is not a right or a wrong language.

Lower-Case Greek Alphabet							
α	alpha	ε	epsilon	ι	iota	ν	nu	ϱ	rho	φ	phi
β	beta	ζ	zeta	κ	kappa	ξ	xi	σ	sigma	χ	chi
γ	gamma	η	eta	λ	lambda	o	omicron	τ	tau	ψ	psi
δ	delta	ϑ	theta	μ	mu	π	pi	υ	upsilon	ω	omega

In the eighteenth century, Messier cataloged 103 nebulae which were later extended to 110 objects. In a few cases his description is not clear, so that some people disagree with the generally accepted identification. Messier objects are designated by an "M". A much more complete list of nonstellar objects is the *New General Catalogue* (NGC) with the *Index Catalogue* (IC). NGC objects are labeled by the number alone, while *Index Catalogue* objects start with "IC". All the nebulae in this atlas are listed on p. 122.

Resolution

The eye has a resolution of $5' = 300''$: it can resolve double stars of $5'$ separation or more. Very good eyes can resolve closer binaries, like ε Lyrae, of $3.5'$ separation. Double stars with a very faint companion are more difficult. When observing

Explanatory Notes _____

with binoculars and telescopes, the resolution increases to 300″/power, assuming perfect optics. This equation yields the resolution data listed in Table 2 for six instruments. In the catalog of binaries, the visibility is indicated by a dice symbol showing how many of these six instruments each binary is accessible to. The binaries observable with each instrument are thus easily identified.

The resolution is also limited by the aperture and is in the best case 120″/aperture in mm (5″/aperture in in.). The resolution of the eye and the telescope match each other if the magnification is 2.5 times the aperture in mm (60 × aperture in in.). This is the highest efficient power. It can be used for binaries under good conditions. On the other hand, most nebulae require a much lower power.

Unfortunately, most standardly equipped telescopes come with high, and completely useless, magnifications, while the so-important low powers, with large fields of view, are missing. A long focal-length eyepiece can easily close this gap. The useful standard magnification is about ten times lower than the highest efficient magnification, that is aperture/4 (6 × aperture in in.). Many binoculars are optimal in this respect and easier to use than a telescope where such a magnification is missing. Many manufacturers like to save money on another part of the telescope as well: the finder. Many finders are made for long searching rather than quick finding. A good finder should have at least a 50 mm aperture and a 5° field of view. The purchase of a good finder can change frustrating searching into exciting observing. Note also that observing with binoculars is much more enjoyable if they are mounted on a tripod.

Table 2: Limiting magnitude and resolving power of six instruments under good conditions (dark sky, steady air, high in the sky, good optics). For nebulae one should decrease the limiting magnitude by one. For each instrucment, the approximate true field of view and the true size of an apparently lunar-size object is also listed. The last column lists the number of observable binaries of the catalog. The visibility of binaries is classified into the six instrument classes by a dice symbol.

Instrument	Power	Aper-ture	Limit. mag.	Field of view	Lunar size	Reso-lution	# of binaries Vis.	# of binaries naries
unaided eye	1×	6 mm	6	120°	30′	300″	⚅	–
opera glasses	3×	20 mm	8	15°	10′	100″	≥⚄	36
finder	6×	30 mm	9	7°	5′	50″	≥⚃	56
binoculars	12×	50 mm	10	4°	2′.5	25″	≥⚂	93
guide scope	60×	75 mm	11	50′	30″	5″	≥⚁	169
telescope {	35× 350×	150 mm	13	80′ 8′	50″ 5″	8″ 0″.8	≥⚀	248

Explanatory Notes

Adaption of the Eye

The human retina has cones and rods. The rods are the sensitive detectors necessary for the observation of nebulae. They are concentrated toward the edge of our field of view. Therefore, the experienced observer looks somewhat away from and not directly at faint nebulae in order to make them detectable. This important observing technique is called indirect vision. When the rods have been blinded, even only for a moment, they need some 30 minutes to regain their full sensitivity. Since rods are sensitive to blue and green light, but not to red light, a deep-sky observer needs a red flash light. This way, one can read star charts without losing the adaption.

Magnitude

Brightness in astronomy is measured in (stellar) magnitudes, abbreviated mag. and sometimes denoted by a superscript ^{m}. The unaided eye can see stars to approximately magnitude 6, depending on sky conditions. Binoculars and telescopes reach to much fainter stars (see Table 2). The main star charts in this atlas represent the naked-eye view (limiting magnitude 6), while the round enlarged sections (finder charts) match the view in a finder or small binoculars (limiting magnitude 9).

Preceding the magnitude entry in the catalog is a small black dot indicating the printed size of the star in the main star chart. This simplifies the comparison between catalog and star chart. The brightest stars in the catalog are thus obvious.

Magnitudes of stars are accurately known. They are listed with one decimal. On the other hand, nebulae do not have well-defined circumferences. Thus their magnitude is very dependent on the area regarded as part of the nebula. It is not surprising that the magnitude of a nebula may differ by a full magnitude in different sources. Therefore, nebular magnitudes are given here to half magnitudes as was done in the New General Catalogue (NGC) more than one hundred years ago. The size and magnitude of a nebula refer in this atlas to what can be seen under a very dark sky. Under less-favorable conditions nebulae will appear fainter and smaller, while professional equipment can trace them further out. Table 3 lists magnitude systems for theoretically interested readers. For practical purposes, the difference of a few tenths of a magnitude between different magnitude systems is not noticeable and thus negligible.

When you actually observe nebulae, the total magnitude is often not as important as the surface brightness or brightness density per square arc-minute. (Both the total magnitude and the surface brightness are listed in the catalog under "v-Mag.") Nebulae with a high surface brightness ($10/\square'$) allow high power (magnification). Thus they can be observed in moonlight or artificial light pollution. Fainter

Explanatory Notes

Table 3: Magnitude Systems.

V-magnitude	V in UBV-system, corresponds to the spectral sensitivity of the eye with direct vision, appropriate parameter for bright stars.
v-magnitude	visual, corresponds to the spectral sensitivity of the eye with indirect vision at night, appropriate parameter for nebulae.
relation	$v = V + (B-V)/3$ for stars
	$v \approx V - 1$ for planetary and diffuse emission nebulae
	$v \approx V$ for open star clusters and reflection nebulae
	$v \approx V + 0.3$ for globular star clusters and galaxies

objects ($12/\square'$) require a low power and dark sky. To find nebulae with a low surface brightness ($14/\square'$) is a challenge, sometimes even to the experienced observer. Those nebula may be visible to the unaided eye in perfect conditions while a search may be hopeless with slight light pollution, even for a large telescope. In the visibility column of the catalog, dice symbols with open circles are a warning sign of low surface brightness.

Table 4: Surface brightness of the background sky as a function of artificial light pollution or moonlight. The sky appears darker through a nebular filter which only helps for specific nebulae. The right part of the table lists the number of observable nebulae of the catalog, restricted by the instrument size and by the condition that the surface brightness of the nebula must at least match the surface brightness of the background. The instrument class necessary to recognize a nebula as such is given by a dice symbol, listed for each nebula in the catalog. Nebulae with low surface brightness ($13-14/\square'$) have a dice symbol with open circles.

Artificial or natural illumination: Location	Moon	Sky background mag.	3×20 Unaided eye ⚅	6×30 Opera glasses ≥⚅	12×50 Fin- der ≥⚅	75mm Binoc- ulars ≥⚅	150mm Guide scope ≥⚅	Tele- scope ≥⚅	Vis.
large city	Full ○	$10/\square'$	5	10	12	14	25	32	
small city	¾ ○	11	18	40	53	64	79	97	⚅–⊡
suburb	Half ◗	12	24	63	86	112	148	170	
field	¼ ◗	13	34	83	116	149	198	232	
mountains	New Moon	14	35	89	123	161	216	250	⚅–⊡

The surface brightness might vary across the nebula, so that bright stars in a cluster or a galaxy core can be more easily observed, while the outer nebular regions might be much more difficult to see. Although knowledge of the surface brightness is very valuable for the deep-sky observer, it is not listed in most atlases or catalogs.

These data together with Table 4 can be the basis for selecting objects for an observing session. For example, an observer with binoculars on a field far from city lights and without moonlight will choose nebulae with a visibility of more than 3 and a surface brightness over 13/□' (a dice symbol with at least four filled dots). An experienced observer may wish to check out the limits (visibility of 3 or surface brightness of 13/□').

Color

Only the very brightest stars reveal their color to the unaided eye. With binoculars, a few hundred stars appear colored, namely the stars down to magnitude 4. In a telescope, stars of magnitude 7 and the brightest nebulae show their colors.

Colors of stars are measured by their B–V color index listed in the catalog. For negative color indices, the zero before the decimal point is omitted. The scale goes from blue with a slightly negative color index via white (B–V = 0.5) to yellow with a color index of 1–2. There are very few stars with a color index exceeding 2 shining orange-red; they are so faint that their colors appear only in a telescope. For comparison, the blue color of the clear daytime sky has a color index of −0.3 while the yellow light of a standard light bulb has a color index of about 1.5. A red giant with a color index of 1.5 is not red but as yellow as a light bulb. This can be easily verified by watching Betelgeuse or Antares through a telescope. For this purpose, it is best to put the bright star slightly out of focus since the human color recognition is more sensitive for extended objects than for point sources.

Color pictures show yellow stars and blue-green nebulae in a red color. Color emulsions have their color balance adjusted for daylight but not for the light of cool stars or emission nebulae. Color emulsions show the universe in false color, a different, fascinating view from the view to the human eye. The false color makes many astronomical pictures more impressive.

The color of a star is directly related to its surface temperature. Stars with white heat are hotter than those with red heat. Still hotter stars glow bluish. In the catalog, a thermometer symbol follows the entry of the color index. This symbol deserves special attention for binary stars since binaries with components of different temperatures show impressive color contrasts.

Explanatory Notes _____

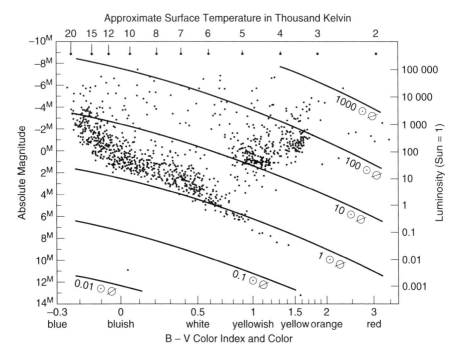

Figure 5: Color-magnitude diagram of all stars of this catalog. Most stars seem to be more luminous than our sun, but this is only a selection effect. Actually, more than 80 percent of the stars are less luminous than our sun. The curves indicate stellar diameters in units of the solar diameter (1.4 million km = 865 000 miles).

Luminosity

The real luminosity of a star is indicated by its absolute magnitude; this is the magnitude from a standard distance of 10 pc = 32.6 light years. It is listed in the catalog (column "Abs."). For variable stars, only the maximum absolute magnitude is listed. For comparison, our sun has an absolute magnitude of $4^{M}\!.8$. The most luminous stars with an absolute magnitude of -8^{M} shine 100 000 times brighter than our sun when viewed from the same distance. On the other end, Barnard's Star (chart E17) with an absolute magnitude of 13^{M} does not even have one thousandth of the solar luminosity.

Table 5: Classification of variable stars.

Intrinsic Variable Stars	Typical light curve	#
irregular: ⎱ giant and supergiant pulsating stars		18
semiregular: ⎰		23
Mira type: long period, large amplitude, light curve changing	⟋‾⟍	14
Cepheid: short period, quite regular, named after δ Cephei	⟋‾⟍	10
Eclipsing Binaries (very regular)		
Algol type: short minima, long constant maximum light	‾‾V‾‾	6
Beta Lyrae (β Lyr) type: almost constantly varying	⌒V⌒	9

The size of a star is determined by its absolute magnitude and color (Fig. 5). Stars at least about 10 times larger than the sun are giants. Those more than 10 times smaller than the sun are white dwarfs. The best observable white dwarf is listed on p. 54, bottom.

Binaries

If the separation between two stars is less than the resolution of the unaided eye (about $5' = 300''$) they appear as one star to the unaided eye. In that case, the catalog of stars contains only one entry with the combined light of both components. It is thus consistent with the appearance of the sky to the unaided eye. A symbol of two stars following the magnitude entry indicates that this star is a double star or binary. Both components are then listed with their magnitudes and colors in the catalog of binaries. Also listed is the separation between both components. The relative position angle between both components is shown graphically with north to the top and west to the right consistent with the orientation in the star charts.

Binaries which hardly move between 2000 and 2015 have one entry for separation and position angle. Binaries with apparent motion are listed for the year 2000 (first entry, abbreviated '0) and further years until 2015. To be exact, the separation and position angle refer to the beginning of a given year, but this really only matters for the fastest binaries. For a few very fast binaries, the motion of the companion relative to the primary is shown graphically between 1995 and 2020 with north to the top as well.

For triple stars with comparable separations, the third component is listed in a second row. For triple stars with very different separations, the close pair is listed in the second row. The first row then shows the appearance in a small instrument where the close pair cannot be separated.

Explanatory Notes _____

There are two types of binaries: intrinsic binaries and optical binaries. They are intrinsic if the stars are close to each other in space. They are optical if they are along our line of sight but at different distances. In 1997 the satellite Hipparcos revealed for many binaries their type due to its accurate distance measurements. Intrinsic binaries have one entry for the distance in the catalog while optical binaries have both distances given with the distance to the brighter star listed first. The type of the binary is also revealed in the catalog of binaries since separations of optical binaries are rounded to full arc-seconds. Some intrinsic binaries are observable with the unaided eye. Each component is then listed in the catalog of stars but not in the catalog of binaries.

Variable Stars

There are two main groups of variables: eclipsing binaries (two stars occult each other) and intrinsic variables (physically changing stars). Eclipsing binaries can be divided into Algol-type and β Lyrae-type stars. Algol-type variables shine mostly at constant, maximum light. Their brightness drops steeply when the stars are eclipsing each other. β Lyrae stars constantly change their brightness. The two stars are so close that they are elongated by gravitational interaction. Sometimes we look at their narrow side, sometimes at their wide side. Intrinsic variables are divided into irregular, semiregular, δ Cephei-type stars or Cepheids (short period), and Mira-type stars (long period), and novae which are not listed in this atlas. This is still only a rough classification: there are more than 50 subtypes of variable stars.

For variable-star observers, the tables list important information. The time of a maximum or minimum (usually the first one after J.D. 2451200) is listed as the Julian Date under "Max." or "Min.". The Julian Date for the beginning of each month is listed on p. 120. Further maxima and minima can be easily calculated by adding a multiple of the given period.

For Algol stars, the duration of the eclipse is given. This is the duration of the magnitude drop and increase. The minimum is centered within this duration. Many intrinsic variable stars have asymmetric light curves, the brightness drop is slower than the increase. A quantified description of the asymmetry is sometimes given. For example, "Min = Max + 10" means that the minimum occurs regularly 10 days after the maximum.

Eclipsing binaries have a secondary minimum half way between successive primary minima, although this is barely noticeable for many variables. Many intrinsic variables display a varying amplitude from period to period. For example, Mira can get as bright as magnitude 2 but reaches only magnitude 4 during other maxima. The catalog of stars lists the average amplitude while the key word "Extrema" gives their extreme values ever recorded.

Nebulae

Most nebulae belong to one of the five groups of planetaries, diffuse nebulae, open and globular star clusters, and galaxies.

Planetary nebulae: These are called "planetaries" because in a telescope they look like small, greenish disks, just like Uranus. They are gaseous nebulae and consist of the outermost shell of a hot central star, blown into space many thousands of years ago. Most planetaries appear stellar in binoculars. Only at high power do they reveal their shapes. Rings and disks are the most common shapes. Planetary nebulae are more conspicuous through a green filter, or rather a nebular filter. This transmits the green or blue-green light of the nebula, but absorbs most of the other parts of the spectrum, thereby increasing the contrast between nebula and background. The colors of nebulae are barely visible in amateurs' telescopes. But in very large telescopes planetaries shine intensely green or bluish-green.

Diffuse nebulae: These consist of gas and dust. Usually one finds them within a young open cluster where new stars are forming from their gas and dust. They are called emission nebulae if most of the light is gaseous emission. Their color, and the use of nebular filters, is the same as for planetaries. Filamentary supernova remnants also emit a similar spectrum. They do not tell the story of the birth of stars, but rather of the end of a star's life. Diffuse nebulae are called reflection nebulae if most of the light is reflected or scattered light from a star by interstellar dust. They are more difficult to observe since their contrast cannot be enhanced by the use of nebular filters.

Open star clusters: Open clusters might appear nebulous to the naked eye or in small binoculars. But in a telescope they are always resolved into individual stars. Many open clusters are very young compared to our solar system. Gas and dust forming new stars are often associated with them. Rich open clusters consist of more than a hundred stars. Clusters poor in stars (less than 50 stars) are usually inconspicuous.

Globular star clusters: Globular clusters are so distant that individual stars only become visible in telescopes. In this case the cluster is said to be resolved. In binoculars and partially still in telescopes, the hundreds of thousands of stars appear as a nearly circular glow. The concentration towards the center is measured on a scale from I (extreme concentration) to XII (very smooth).

Galaxies: These are systems of stars like our own Milky Way. Large telescopes reveal their different shapes. Elliptical galaxies (E) appear as a featureless, elliptical glow. Lenticular galaxies (S0) look similar, but they contain dust clouds which show up as dark patches. Spiral galaxies come in a wide variety. Some are similar to lenticulars, but have faint spiral arms outside their bulge (Sa). The other extreme is a spiral with very long arms but no indication of a central bulge (Sd). Irregular spiral arms (Sm) mark the transition to the irregular galaxies (Ir) which do not fit

Explanatory Notes

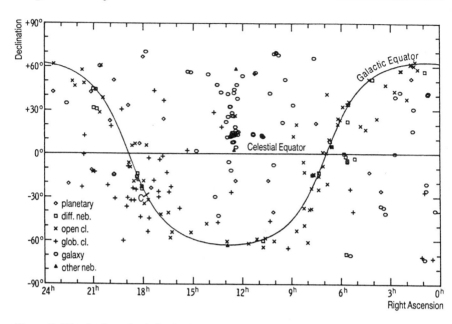

Figure 6: Distribution of nebulae in the sky. Planetaries, diffuse nebulae, and open clusters lie near the central plane of the Milky Way, the galactic equator. Globular clusters are also found far away from this plane, but are concentrated toward the center C of the Milky Way. Galaxies avoid the galactic equator and are densest in the Virgo Cluster (near center of figure). In the plane of the Milky Way, interstellar absorption by dust limits the visibility to approximately 10 000 light-years. Looking out of the plane, we have a clear view millions of light-years deep into the universe. Similarly, on a hazy day, the visibility towards the horizon may be very limited, yet the sun, some 150 million kilometers away, can be seen clearly high up in the sky

into any of the other groups. There are more differentiated classification schemes, which also distinguish barred spirals from standard spirals. Finer subdivisions of galaxies are not easily observable in amateurs' telescopes.

Oblateness: Elliptical galaxies are classified according to their oblateness from E0 (circular) to E7 (very elongated). The oblateness is also important for the observation of other nebulae. For example, an apparently circular spiral galaxy is best suited for tracing spiral arms while an extremely elongated spiral galaxy shows most likely dust in the central plane. In the catalog, the oblateness of each nebula is graphically shown in the column "Shape," from a circle all the way to a thin line.

The descriptions of the 250 nebulae are based on observations by the author, who has observed every one with a telescope of 150 mm aperture under dark sky.

Table 6: Classification of nebulae. Abbreviations listed under "Type" and "Shape" as well as the oblateness symbols are used in the catalog of nebulae.

Type, Shape	Description	#	Type, Shape	Description	#
PN	**planetary nebula**		**GC**	**globular cluster**	
R	ring shaped	7	I–IV	core bright, concentrated	14
D	disk shaped	10	V–VIII	medium concentration	17
A	anomalous, irregular	4	IX–XII	uniform, without core	12
DN	**diffuse nebula**		**Glx**	**galaxy**	
Em	emission nebula	16	E	elliptical galaxy	12
Re	reflection nebula	4	0–7	oblateness	
Fi	filamentary supernova	3	S0	lenticular galaxy	10
	remnant		Sa–m	spiral galaxy:	
			a	large core, hardly arms	9
OC	**open cluster**		b	medium core, short arms	15
r	rich in stars	21	c	small core, medium arms	14
m	medium number of stars	36	d	no core, long arms	12
p	poor in stars	21	m	irregular spiral arms	4
n	visible nebulosity	11	Ir	irregular galaxy	5

oblateness $= 10\,(a - b)\,/\,a$	oblateness:	0	1	2	3	4	5	6	7	8	9	
a **major axis**, b **minor axis**	shape:	O	O	O	O	O	O	O	O	I	I	

Further information comes from the much larger volume "Observing Handbook and Catalogue of Deep-Sky Objects." Descriptions for the view in binoculars refer to 12×50 mm binoculars. Of course, the visible details are very dependent on many parameters, especially the darkness of the sky, so that they can only give a rough idea about what to expect to see.

Among the stars there are variable stars. On the other hand, nebulae do not change their light and shape, with two exceptions: Hubble's Variable Nebula (see p. 64) and the expanding light echo of the supernova 1987A near the Tarantula Nebula (see p. 102). It is not known how bright it will develop within the coming years. Time will tell.

Further Reading

Sky Atlas 2000.0 by Wil Tirion. Cambridge University Press and Sky Publishing Corporation, 1981.

Explanatory Notes

This large-format atlas with 43,000 stars to visual magnitude 8.0 plus 2,500 deep-sky objects is the ideal supplement for the advanced observer.

Sky Catalogue 2000.0 (2 vols.) edited by Alan Hirshfeld and Roger W. Sinnott. Cambridge University Press and Sky Publishing Corporation, 1982 (Vol. 1), 1985 (Vol. 2).
Data and notes on nearly all of the stars and objects of *Sky Atlas 2000.0* are given in this catalog for the advanced observer.

Catalog Headings

NEBULA Designation, the first column lists the number from the New General Catalogue (NGC); IC: Index Catalogue, M: Messier.

Position   Constellation (see pp. 126, 127); dot and circle in the rectangle show the location in the main star chart and enlargements.

v-Mag.   Total visual magnitude (first entry) and surface brightness per square arc-minute (second entry, mag./$\square'$).

Size   Apparent diameter of a nebula in arc-minutes ($'$).

Shape   Classification according to the appearance in a telescope (see Table p. 19). Preceeding is an oval showing the elongation.

Type   Classification into PN: planetary nebulae, DN: diffuse nebulae, OC: open clusters, GC: globular clusters, and Glx: galaxies.

Vis.   Visibility from telescope only ⊡ to unaided eye ⊞. Low surface brightness ◌–▦ requires dark sky (see Table 4, p. 12).

Dist.   Distance in light years (M: million).

R.A.   Right ascension for the equinox 2000.0, in hours and minutes.

Dec.   Declination for the equinox 2000.0, in degrees.

STAR Designation usually consists of the Flamsteed number and/or the Bayer Greek letter followed by the constellation.

Position   Indicates the location in the star chart.

V-Mag.   Magnitude V in the UBV-system, combined magnitude for binaries ⚹. A preceeding dot shows the size in the main star chart.

B−V   Spectral color index, from bluish (< 0) to orange-red (> 2).

Te.   Symbol for the surface temperature of a star, from cool, yellow stars ↓ to hot, bluish stars ↓.

Abs.   Absolute magnitude, the V-mag. from a distance of 32.6 ly.

Name   Historic name of a star; names in use are printed in bold.

Dist.   Distance in light years.

R.A.   Right ascension for the equinox 2000.0, in hours and minutes.

Dec.   Declination for the equinox 2000.0, in degrees.

BINARY The same designation as in the catalog of stars.

Position   Indicates the location in the star chart.

V-Mag.   Magnitudes V in the UBV-system for both components.

B−V   Spectral color indeies in the UBV-system for both components.

Te.   Symbols for the surface temperatures of both components.

Sep.   Separation in arc-seconds between both components.

PA   Relative position angle between both components (north up).

Vis.   From telescope only ⊡ to unaided eye ⊞ (see Table p. 10).

VARIABLE STAR   The first line gives the designation, the position, the size in the main star chart, and the light curve (see Table p. 16).

Period   Duration of the periodicity listed in days (d).

Max./Min.   Julian Day of maximum/minimum brightness (see pp. 118, 119).

Extrema   Extreme magnitudes ever observed.

Eclipse   Duration of eclipse for eclipsing variable stars.

2nd min.   Magnitude at secondary minimum for eclipsing variable stars.

Arrangement of Charts (see also p. 132)

	Fall	Summer	Spring	Winter	Fall
60°	N22	N16 Northern Sky	N10 N8	N2	
	N24 N20	N18 N14	N12	N6 N4	N0
30°					
	E23 E21	E19 E15	E13 E11 ···· E7	E3 ···· E1	
0°		Equator	E14 ···· E9	E5	
	E24 ·····	E17		E8	
	E22 Ecliptic····	E12 E10	E4 E2	E0	
−30°	E20 E18 E16		E6		
	S21	S18 S15 S12	S6		
−60°	S24	Southern Sky	S9	S3	S0
	24ʰ 21ʰ	18ʰ 15ʰ	12ʰ 9ʰ	6ʰ 3ʰ	0ʰ

STAR		Position	V-Mag.		B−V	Te.	Abs.	Name	Dist.	R.A.	Dec.
48	Cas	• •	4.5	☆	0.2	↓	2^M	....	117 ly	$2^h02^m.0$	70°.91
50	Cas	• •	4.0		0.0	↓	0	..... ⌈Star⌉	160	2 03.4	72.42
1 α	UMi	• ●	2.0		0.6	↓	−4	Polaris, North	430	2 31.8	89.26
78	Cam	• •	4.8	☆	0.0	↓	0	also 32 Cam	320	12 49.2	83.41
5	UMi	• •	4.3		1.4	↓	−1	....	350	14 27.5	75.70
7 β	UMi	• ●	2.1		1.5	↓	−1	.. Kochab ..	125	14 50.7	74.16
13 γ	UMi	• ●	3.0		0.1	↓	−3	.. Pherkad ..	460	15 20.7	71.83
16 ζ	UMi	• •	4.3		0.0	↓	−1	....	370	15 44.1	77.79
21 η	UMi	• •	5.0		0.4	↓	3	....	97	16 17.5	75.76
22 ε	UMi	• •	4.2		0.9	↓	−1	....	350	16 46.0	82.04
23 δ	UMi	• •	4.4		0.0	↓	1	....	180	17 32.2	86.59
41,40	Dra	• •	5.1	☆	0.5	↓	1	....	170	18 00.1	80.00
75	Dra	• •	5.1	☆	1.0	↓	−1	....	500	20 28.0	81.43

BINARY		Position	V-Mag.		B−V		Te.	Sep.	PA	Vis.
48	Cas	• •	4.7	6.7	0.1	0.4	↓↓'0	0".8	••	⊙
								2005 0.8	••	⊙
								2010 0.7	••	◯
								2015 0.6	••	◯
78	Cam	• •	5.3	5.8	0.0	0.0	↓↓	21.5	•	⊙
41,40	Dra	• •	5.7	6.0	0.5	0.5	↓↓	18.8	•.	⊙
75	Dra	• •	5.4	6.6	1.0	1.0	↓↓	196.7	••	⊠

(2020, 1995 diagram for 48 Cas)

Polaris	Coordinates	
1900.0	$1^h22^m.6$	88°.77
1950.0	1 48.8	89.03
1990.0	2 21.2	89.22
2000.0	2 31.8	89.26
2010.0	2 43.7	89.31
2020.0	2 57.1	89.35
2050.0	3 48.3	89.45

Stellar diameters and scale in star charts

in round enlarged sections

Stellar magnitudes in tenths of a magnitude (20 = 2ᵐ0, etc.)

NEBULA		Position		v-Mag.	Size	Shape	Type	Vis.	Dist.	R.A.	Dec.
205	M110	And	[◉]	8½	12/□′ 10′	0 E5	Glx	[☺]	3 Mly	0ʰ40ᵐ.4	41°.69
221	M32	And	[◉]	8½	11	3.5 O E2	Glx	[☺]	3 M	0 42.7	40.87
224	M31	And	[◉]	4	13	150 ⟦ Sb	Glx	[◉◉]	3 M	0 42.7	41.27
598	M33	Tri	[◉]	6	14	50 0 Sd	Glx	[※]	3 M	1 33.9	30.66
650	M76	Per	[◔]	10	11	2.5 0 A	PN	[•]	4 000	1 42.4	51.57
752	.....	And	[◔]	6	14	50 O m	OC	[※]	1 500	1 57.8	37.68
891	.....	And	[◔]	10½	13	10 │ Sb	Glx	[◦]	40 M	2 22.6	42.33

205	M110	Companion galaxy of the Andromeda Galaxy, slightly asymmetric.
221	M32	Companion of the Andromeda Galaxy, almost stellar in binoculars.
224	M31	**Andromeda Galaxy**, nearest large galaxy, physically comparable with our Milky Way, bright prominent core, dust lanes west of the core, outer spiral arms and great size visible only under dark sky.
598	M33	**Triangulum Galaxy**, dark sky and low power essential, elongated glow in binoculars without a bright core; a telescope shows two or three spiral arms with emission nebulae and stellar associations.
650	M76	**Little Dumbbell**, irregular shape, consists of NGC 650 and 651.
752	.....	Difficult object with unaided eye, nicely resolved in binoculars.
891	.....	Faint edge-on galaxy, very elongated shape distinct in a telescope.

STAR				Position	V-Mag.	B−V	Te.	Abs.	Name	Dist.	R.A.	Dec.
21	α	And	[▪]	●	2.1	0.0	↓	0ᴹ	Alpheratz, Sirrah	98 ly	0ʰ08ᵐ.4	29°.09
31	δ	And	[▪]	●	3.3	1.3	↓	1	. . . .	102	0 39.3	30.86
35	ν	And	[◉]	•	4.5	−.1	↓	−2	. . . .	650	0 49.8	41.08
37	μ	And	[◉]	•	3.9	0.1	↓	1	. . . .	140	0 56.8	38.50
43	β	And	[▪]	●	2.1	1.6	↓	−2	. . **Mirach** . .	200	1 09.7	35.62
50	υ	And	[▪]	•	4.1	0.5	↓	3	. . . .	44	1 36.8	41.41
51		And	[◔]	•	3.6	1.3	↓	0	. . . .	180	1 38.0	48.63
	φ	Per	[◔]	•	4.0	−.1	↓	−3	. . . .	800	1 43.7	50.69
2	α	Tri	[◉]	•	3.4	0.5	↓	2	. Elmuthalleth .	64	1 53.1	29.58
56		And	[◉]	•	5.0	✴ 1.3	↓	−2	. . . .	320, 900	1 56.0	37.26
57	γ	And	[◉]	●	2.1	✴ 1.2	↓	−3	. . **Alamak** . .	370	2 03.9	42.33
4	β	Tri	[◉]	●	3.0	0.1	↓	0	. . . .	125	2 09.5	34.99
59		And	[◉]	·	5.6	✴ 0.0	↓	1	. . . .	300	2 10.9	39.04
6	ι	Tri	[▪]	·	4.9	✴ 0.8	↓	0	. . . .	300	2 12.4	30.30
9	γ	Tri	[▪]	•	4.0	0.0	↓	1	. . . .	120	2 17.3	33.85
15		Tri	[▪]	·	5.1	✴ 1.1	↓	−2	. . . .	1 000	2 35.8	34.70
R		Tri	[▪]	·	6.0–10	1.3	↓	−2	. . . .	1 000	2 37.0	34.26

BINARY		Position		V-Mag.		B−V		Te.	Sep.	PA	Vis.
56	And	[◉]	·	5.7	5.9	1.1	1.6	↓↓	201″	•·	[⊡]
57	γ And	[◉]	●	2.2	4.9	1.4	0.0	↓↓	9.6	•·	[⊡]
59	And	[◉]	·	6.1	6.8	0.0	0.1	↓↓	16.7	•·	[⊡]
6	ι Tri	[▪]	·	5.2	6.7	0.8	0.5	↓↓	3.9	•·	[⊡]
15	Tri	[▪]	·	5.4	6.7	1.6	0.2	↓↓	142.2	⁖	[⊡]

VARIABLE	STAR
R Tri [▪] · ⌇‿⌐	
Period	267 d
Max.	2451368
Min. Max. + 150	
Extrema	5.4–12.6

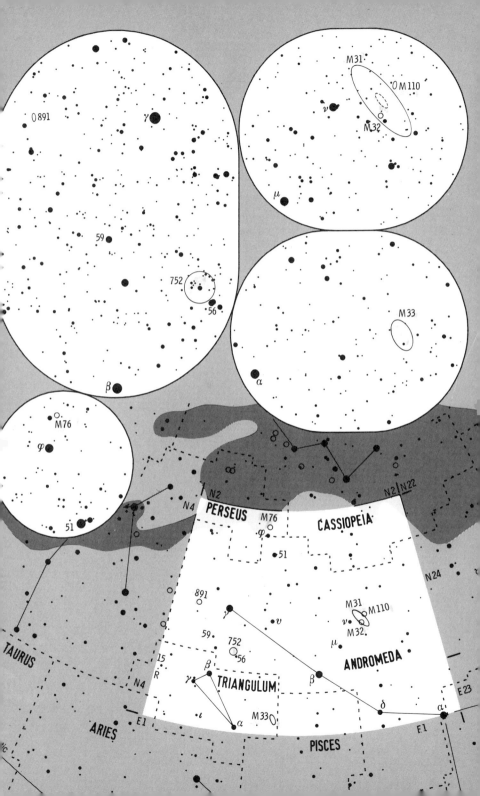

NEBULA	Position		v-Mag.	Size		Shape	Type	Vis.	Dist.	R.A.	Dec.
281	Cas	⊙	7½	14/□′	20′	O Em	DN	⚲	7000 ly	0ʰ52ᵐ.8	56°.60
457	Cas	⊙	6½	11	12	O r	OC	⚃	5000	1 19.1	58.33
559	Cas	⊙	8	11	6	0 m	OC	⚀	4000	1 29.5	63.30
581 M103	Cas	⊙	7½	11	6	O p	OC	⚄	7000	1 33.2	60.70
654	Cas	⊙	7	10	5	O m	OC	⚃	7000	1 44.1	61.88
663	Cas	⊙	7	12	15	0 m	OC	⚄	7000	1 46.0	61.27
869	Per	⚲	4	11	25	O r	OC	⚅	8000	2 19.0	57.13
884	Per	⚲	4	11	25	O r	OC	⚅	8000	2 22.4	57.12

281	Faint in binoculars, interesting in a telescope with nebula filter.
457	Well resolved in binoculars, remarkable stellar pattern in a telescope.
559	Looks like a faint oval nebula in binoculars, resolved in a telescope.
581 M103	Resolved in binoculars, hardly better in a telescope, a star mag. 7.3.
654	Contains many faint stars, therefore mostly nebulous appearance.
663	Excellent even in binoculars, many individual stars in a telescope, quite irregular shape, contains two regions with many faint stars.
869	h Persei ⎱ **Double Cluster, h and chi Persei**, easily visible with
884	χ Persei ⎰ unaided eye as an elongated nebula, splendid view in binoculars, still better in a telescope at low power, each cluster displays some 60 stars but actually contains approximately 300 stars.

STAR	Position		V-Mag.	B–V	Te.	Abs.	Name	Dist.	R.A.	Dec.
11 β Cas	⚫	●	2.3	0.4	↓	1ᴹ	.. Chaph ..	55 ly	0ʰ09ᵐ.2	59°.15
14 λ Cas	⚫	•	4.7	−.1	↓	0	....	340	0 31.8	54.52
17 ζ Cas	⚫	●	3.7	−.2	↓	−3	....	600	0 37.0	53.90
18 α Cas	⊙	●	2.2	1.2	↓	−2	.. Schedir ..	240	0 40.5	56.54
24 η Cas	⊙	•	3.4 ✳	0.6	↓	5	....	19.4	0 49.1	57.82
27 γ Cas	⊙	●	2.2–2.5	−.1	↓	−4	....	600	0 56.7	60.72
34 φ Cas	⊙	•	4.8 ✳	0.6	↓	−7	near NGC 457	5000	1 20.1	58.23
37 δ Cas	⊙	●	2.7	0.2	↓	0	. Ruchbah .	100	1 25.8	60.24
45 ε Cas	⊙	●	3.4	−.1	↓	−2	....	450	1 54.4	63.67
ι Cas	⚫	•	4.5 ✳	0.1	↓	1	....	140	2 29.1	67.40
SU Cas	⚫	·	5.7–6.2	0.7	↓	−3	....	1500	2 52.0	68.89
9 α Cam	⚫	•	4.3	0.0	↓	−7	....	4000	4 54.1	66.34
10 β Cam	⚫	•	4.0 ✳	0.9	↓	−4	....	1000	5 03.4	60.44
11,12 Cam	⚫	•	4.8 ✳	0.2	↓	−2	....	700	5 06.2	58.98

BINARY	Position		V-Mag.		B–V		Te.	Sep.	PA	Vis.
24 η Cas	⊙	●	3.5	7.4	0.6	1.4	↓↓	′0 12″.8	•ʹ	⚀
								2015 13.3	•ʹ	⚀
34 φ Cas	⊙	·	5.0	7.0	0.7	0.4	↓↓	134.1	•.	⚄
ι Cas	⚫	·	4.6	6.9	0.1	0.4	↓↓	′0 2.9	•.	⚀
								2015 3.0	•.	⚀
10 β Cam	⚫	·	4.0	7.4	0.9	0.3	↓↓	83.4	•.	⚋
11,12 Cam	⚫	·	5.2	6.1	−.1	1.1	↓↓	178.5	⦂	⚃

VARIABLE STAR

27 γ Cas	⊙ ● irregular
	Period > 1 d
	Extrema 1.6–3.0
SU Cas	⚫ · ⌒
	Period 1.94931 d
	Max. 2451201.48
	Min. Max. + 1.2

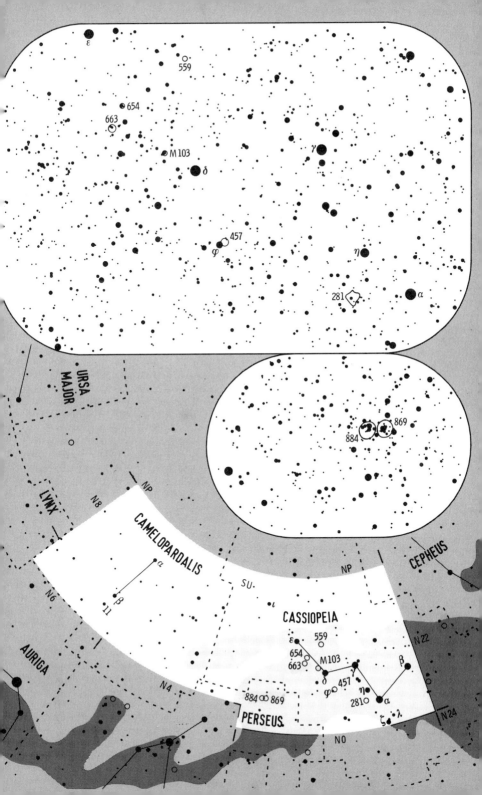

NEBULA	Position	v-Mag.	Size	Shape	Type	Vis.	Dist.	R.A.	Dec.
1023	Per	10	12/□′	5′	SO	Glx		45 Mly	$2^h40^m.4$	39°.06
1039 M34	Per	5½	12	30	○ m	OC		1500	2 42.0	42.78
1245	Per	8½	13	10	○ r	OC		7000	3 14.7	47.27
1491	Per	10	12	3	○ Em	DN		2500	4 03.4	51.32
1528	Per	6½	13	25	○ m	OC		2500	4 15.4	51.23

1023	Faint elliptical nebula with a bright core but without other features.
1039 M34	Very nice cluster for a finder and for binoculars, interesting in a telescope at low power only, some stars seem to be aligned along arms, others make various patterns, distinct central condensation.
1245	Faint nebula in binoculars; individual stars become visible in a telescope but the background remains nebulous due to many faint stars.
1491	At high power with nebula filter well separated from mag. 11.0 star.
1528	Interesting resolved cluster in every telescope, some individual stars are visible even in binoculars, irregular distribution of faint stars.

STAR	Position	V-Mag.	B–V	Te.	Abs.	Name	Dist.	R.A.	Dec.
16	Per	• 4.2	0.3		1^M	. . . .	130 ly	$2^h50^m.6$	38°.32
15 η	Per	• 3.8	1.7		−4	. . . .	1200	2 50.7	55.90
18 τ	Per	• 3.9	0.8		−1	. . . .	250	2 54.3	52.76
23 γ	Per	● 2.9	0.7		−2	. . . .	250	3 04.8	53.51
25 ϱ	Per	● 3.3–3.5	1.6		−2	. . . .	310	3 05.2	38.84
26 β	Per	● 2.1–3.4	0.0		0	. . Algol . .	93	3 08.2	40.96
ι	Per	• 4.1	0.6		4	. . . .	34.5	3 09.1	49.61
27 κ	Per	• 3.8	1.0		1	. . . .	113	3 09.5	44.86
33 α	Per	● 1.8	0.5		−5	. Mirphak .	600	3 24.3	49.86
39 δ	Per	• 3.0	−.1		−3	. . . .	600	3 42.9	47.79
38 o	Per	• 3.8	0.0		−4	. . . Atik . . .	1200	3 44.3	32.29
41 ν	Per	• 3.8	0.4		−3	. . . .	600	3 45.2	42.58
44 ζ	Per	● 2.8	0.1		−5	. . . .	1200	3 54.1	31.88
45 ε	Per	● 2.9	−.2		−3	. . . .	600	3 57.9	40.01
46 ξ	Per	• 4.0	0.0		−4	. . Menkib . .	1500	3 59.0	35.79
48 υ	Per	• 4.0	0.0		−2	. . . .	600	4 08.7	47.71
51 μ	Per	• 4.1	0.9		−3	. . . .	750	4 14.9	48.41
1	Cam	· 5.4	0.1		−5	. . . .	4000	4 32.0	53.91
57	Per	· 5.6	0.3		1	. . . .	220	4 33.4	43.05
2	Cam	· 5.4	0.3		1	. . . .	280	4 40.0	53.47

BINARY	Position	V-Mag.	B–V	Te.	Sep.	PA	Vis.
45 ε	Per	● 2.9	7.5	−.2	0.0	9″.0	
1	Cam	· 5.8	6.9	0.1	0.1	10.3	
57	Per	· 6.1	6.8	0.4	0.2	120.9	
2	Cam	· 5.6	7.4	0.3	0.5	’0	0.7
		2020 ← 1995			2007	0.8	
					2015	0.8	

VARIABLE STAR

25 ϱ Per • semireg.
Extrema 3.3–4.0

26 β Per ● ⌐‾

Period 2.86731 d
Min. 2451201.24
Eclipse 10 hours

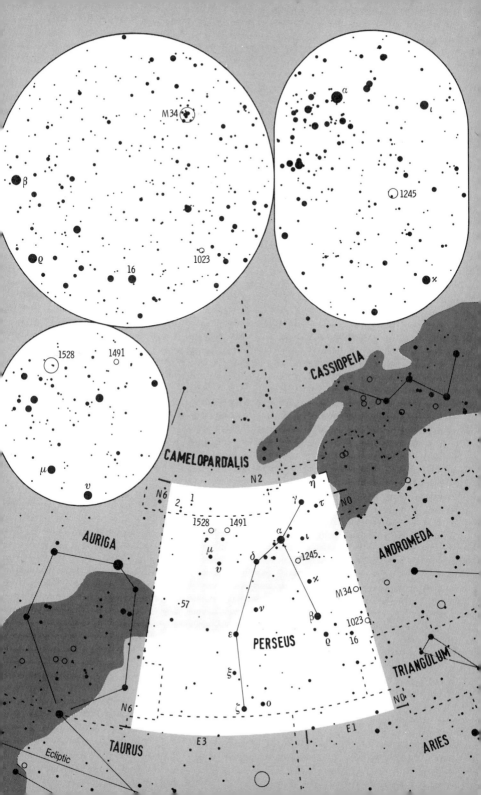

M34

1245

β

ϱ

16

1023

1528   1491

μ

ν

CASSIOPEIA

CAMELOPARDALIS

N2

N6   2   1

1528 ○   1491 ○

μ

ν

AURIGA

57

ν

ε

ξ

ζ

N6

TAURUS

Ecliptic

E3

η

γ

τ

α

δ   ι

1245

χ

β

ϱ   16

PERSEUS

M34

1023

N0

ANDROMEDA

TRIANGULUM

N0

E1

ARIES

NEBULA		Position		v-Mag.	Size	Shape	Type	Vis.	Dist.	R.A.	Dec.
1912	M 38	Aur	○	6½	12/□′	20′	O m	OC	⊠	4000 ly	5ʰ28ᵐ7	35°83
1931	.....	Aur	○	10	11	2.5	O Em	DN	⊙	6000	5 31.4	34.23
1960	M 36	Aur	○	6	12	15	O m	OC	⊠	4000	5 36.1	34.13
2099	M 37	Aur	○	6	12	25	O r	OC	⊠	4000	5 52.4	32.53
2281	.....	Aur	ρ	6	12	20	O p	OC	⊠	2000	6 49.3	41.07

1912	M 38	Partially resolved in binoculars, interesting grouping of faint stars.
1931	.....	Small faint diffuse nebula, imbedded stars visible at high power.
1960	M 36	Some stars resolved in binoculars, about 60 stars in a telescope, aligned along arms, deficiency of faint stars, central condensation.
2099	M 37	Binoculars show a large oval glow, which turns into an amazing number of stars in a telescope, a yellow mag. 9.1 star is centered.
2281	.....	A few bright, irregularly scattered stars in binoculars, oval core.

STAR			Position	V-Mag.	B−V	Te.	Abs.	Name	Dist.	R.A.	Dec.
3	ι	Aur	●	2.7	1.5	↓	−3ᴹ	....	500 ly	4ʰ57ᵐ0	33°17
4	ω	Aur	·	4.9	0.0	↓	1	....	160	4 59.3	37.89
7	ε	Aur	●	3.0–3.8	0.5	↓	−7	....	3000	5 02.0	43.82
8	ζ	Aur	●	3.7–4.0	1.2	↓	−3	....	800	5 02.5	41.08
10	η	Aur	●	3.2	−.2	↓	−1	....	220	5 06.5	41.23
14		Aur	·	4.9	0.2	↓	0	....	270	5 15.4	32.69
13	α	Aur	●	0.1	0.8	↓	0	.. Capella ..	42	5 16.7	46.00
26		Aur	·	5.4	0.4	↓	0	....	450	5 38.6	30.49
32	ν	Aur	●	4.0	1.1	↓	0	....	220	5 51.5	39.15
33	δ	Aur	●	3.7	1.0	↓	1	....	140	5 59.5	54.28
34	β	Aur	●	1.9	0.1	↓	0	. Menkalinan .	82	5 59.5	44.95
37	ϑ	Aur	●	2.6	−.1	↓	−1	....	175	5 59.7	37.21
41		Aur	·	5.8	0.1	↓	1	....	300	6 11.6	48.71
5		Lyn	·	5.1	1.5	↓	−2	......	650, 1500	6 26.8	58.42
48		Aur	·	4.9–5.8	0.7	↓	−4	. RT Aurigae .	2000	6 28.6	30.49
12		Lyn	·	4.8	0.1	↓	1	....	230	6 46.2	59.44
15		Lyn	·	4.4	0.8	↓	1	....	170	6 57.3	58.42
19		Lyn	·	5.3	−.1	↓	−1	....	500	7 22.9	55.29

BINARY			Position	V-Mag.		B−V		Te.	Sep.	PA	Vis.
4	ω	Aur	·	5.0	8.0	0.0	0.5	↓↓	4″9		⊙
14		Aur	·	5.0	7.9	0.2	0.4	↓↓	14.3	•.	⊙
37	ϑ	Aur	●	2.7	7.1	−.1	0.5	↓↓	3.6	•ˑ	⊙
41		Aur	·	6.2	7.0	0.1	0.2	↓↓	7.6		⊙
5		Lyn	·	5.2	7.8	1.5	1.1	↓↓	95	••	⊙
12		Lyn	·	4.9 7.2		0.1	0.3	↓↓	8.9	•ˑ	⊙
				5.4 6.0		0.1	0.1	↓↓ '0	1.8	••	⊙
								2015	1.9	••	⊙
19		Lyn	·	5.4 7.6		−.1	0.0	↓↓	213.5		⊠
				5.8 6.8		−.1	0.0	↓↓	14.8	•ˑ	⊙

VARIABLE	STAR
7 ε Aur	
Min.	July 2010
Eclipse	22 months
8 ζ Aur	
Period	972.2 d
Min.	2451997
Eclipse	40 days
48 RT Aur	
Period	3.7281 d
Max.	2451200.9

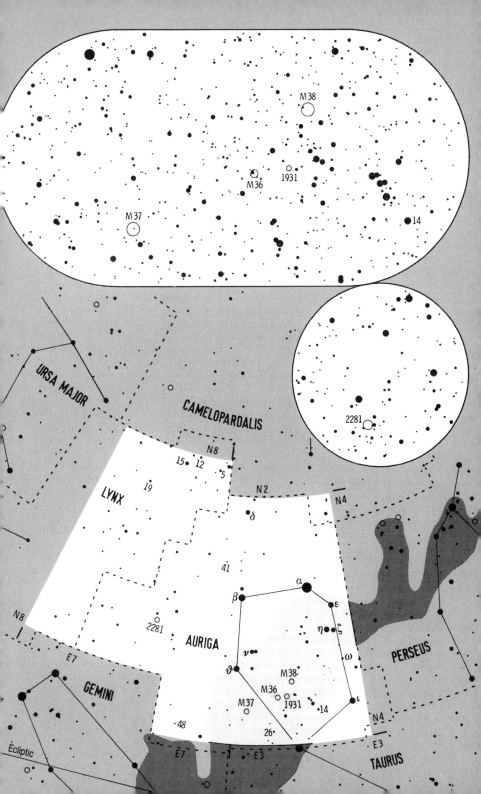

NEBULA	Position	v-Mag.	Size	Shape	Type	Vis.	Dist.	R.A.	Dec.
2403	Cam	8½ 13/□'	12'	0 Sd	Glx		10 M ly	7ʰ36ᵐ.9	65°.60
2683	Lyn	10 12	8	❙ Sb	Glx		20 M	8 52.7	33.42
2841	UMa	9½ 13	7	0 Sb	Glx		35 M	9 22.0	50.98
2976	UMa	10½ 12	4	0 Sc	Glx		13 M	9 47.3	67.91
3031 M81	UMa	7 12	18	0 Sa	Glx		13 M	9 55.6	69.07
3034 M82	UMa	8½ 12	10	0 Ir	Glx		13 M	9 55.9	69.68
3077	UMa	10 12	3	O Ir	Glx		13 M	10 03.4	68.73

2403 ..... Seen well in binoculars, spiral arms barely visible in a telescope.

2683 ..... Faint edge-on galaxy, dust features faintly visible in a telescope.

2841 ..... Small bright nonstellar core within a distinctly elongated nebula.

2976 ..... Companion galaxy of M81, faint elliptical nebula in a telescope.

3031 M81 Central galaxy in a group of galaxies, easily visible in binoculars, bright round central region with stellar core in a telescope, elongated halo; a field of view of 45′ gives a nice view of the pair M81, M82.

3034 M82 Brightest companion of M81, 37′ north of M81, active, almost edge-on galaxy, asymmetric distribution of brightness; a telescope shows wonderful dust features dividing the central area into three parts.

3077 ..... Companion galaxy of M81, featureless nebula with bright core.

STAR		Position	V-Mag.	B–V	Te.	Abs.	Name	Dist.	R.A.	Dec.
31		Lyn	• 4.3	1.5		−1ᴹ		400 ly	8ʰ22ᵐ.8	43°.19
1	o	UMa	• 3.3–3.4	0.9		0		180	8 30.3	60.72
9	ι	UMa	• 3.1	0.2		2 .. Talitha ..		48	8 59.2	48.04
10		UMa	• 4.0	0.5		3 .. in Lynx ..		53	9 00.6	41.78
12	κ	UMa	• 3.6	0.0		−2		400	9 03.6	47.16
38		Lyn	• 3.8 ✶	0.1		1		122	9 18.8	36.80
40	α	Lyn	● 3.1	1.5		−1		220	9 21.1	34.39
41		Lyn	· 5.3 ✶	1.0		1 in Ursa Major		290	9 28.7	45.60
23		UMa	• 3.7	0.4		2		76	9 31.5	63.06
25	ϑ	UMa	● 3.2	0.5		3		44	9 32.9	51.68
24		UMa	· 4.5	0.8		2		105	9 34.5	69.83
29	υ	UMa	• 3.8	0.3		1		115	9 51.0	59.04
31	β	LMi	• 4.2	0.9		1		145	10 27.9	36.71
46		LMi	• 3.8	1.0		1 . also o LMi .		98	10 53.3	34.21

Constellation Boundaries (dashed in star charts): At the time Flamsteed numbered the stars 300 years ago, there were no fixed boundaries between constellations. Not until 1930 were they defined by the International Astronomical Union. Because of the new slightly shifted boundaries, 10 Ursae Majoris and 41 Lyncis are not located within the constellation of their Flamsteed designation.

BINARY	Position	V-Mag.	B–V	Te.	Sep. PA Vis.		VARIABLE STAR
38	Lyn	• 3.9 6.2	0.1 0.5	ⵏ	2.7 •.		1 o UMa • irreg. ?
41	Lyn	· 5.4 7.8	1.0 0.6	ⵏ	71.3 !		Extrema 3.3–3.8

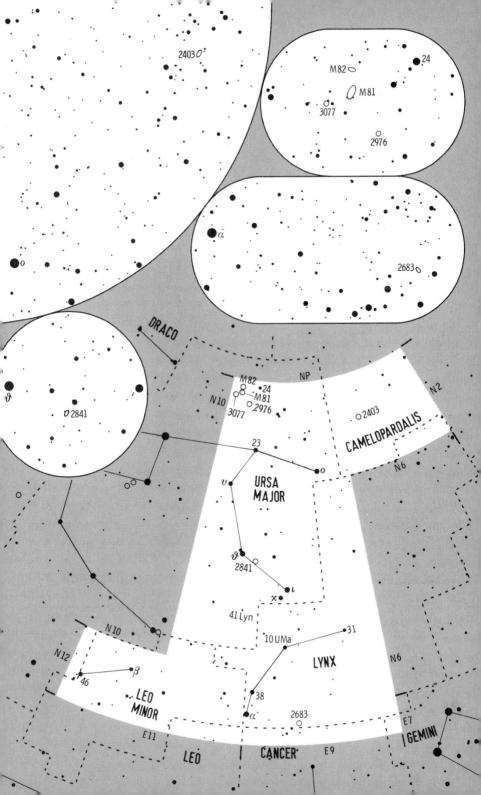

NEBULA		Position	v-Mag.	Size		Shape	Type	Vis.	Dist.	R.A.	Dec.
3184	.....	UMa ⬚	10	13/□′	5′	O Sd	Glx	⊙	35 Mly	10ʰ18ᵐ.3	41°.42
3556	M108	UMa ⬚	10½	13	8	‖ Sd	Glx	⊙	45 M	11 11.5	55.67
3587	M97	UMa ⬚	10	12	3	O D	PN	⊙	2500	11 14.8	55.02
3992	M109	UMa ⬚	10	13	6	0 Sc	Glx	⊙	60 M	11 57.6	53.38
	M40	UMa ⬚	9	(8)	0.9	‖ Dark Neb.		⊙	500	12 22.3	58.08
5457	M101	UMa ⬚	8	14	20	O Sd	Glx	⊙	25 M	14 03.2	54.35

3184 ..... Faint, hardly showing any structure, spiral arms not observable.

3556 M108 Distinct edge-on galaxy, elongated central region, a hint of dust features in a telescope; nice pairing with M97 in 1° field of view.

3587 M97 **Owl Nebula**, circular; both dark eyes are hardly observable.

3992 M109 Very faint Messier object, contains nonstellar central condensation.

M40 Binary: two stars mag. 9.6 and 10.1 in 52″ separation, which exactly matches Messier's description, certainly the correct identification.

5457 M101 **Pinwheel Galaxy**, often just the bright central core is visible, only with darkest sky and lowest power does the enormous size become apparent; spiral arms are hardly visible although a few bright knots are discernable, especially near the southwestern edge, asymmetric.

STAR			Position	V-Mag.	B−V	Te.	Abs.	Name	Dist.	R.A.	Dec.
33	λ	UMa ⬚	●	3.5	0.0	↓	0ᴹ	Tania Borealis	135 ly	10ʰ17ᵐ.1	42°.91
34	μ	UMa ⬚	●	3.1	1.6	↙	−1	Tania Australis	250	10 22.3	41.50
VY		UMa ⬚	·	5.9–6.1	2.4	↙	−2	. . . .	1200	10 45.1	67.41
48	β	UMa ⬚	●	2.3	0.0	↓	0	. . . Merak . . .	80	11 01.8	56.38
50	α	UMa ⬚	●	1.8	1.1	↓	−1	. . . Dubhe . . .	125	11 03.7	61.75
52	ψ	UMa ⬚	·	3.0	1.1	↓	0	. . . .	145	11 09.7	44.50
63	χ	UMa ⬚	·	3.7	1.2	↓	0	. . . .	200	11 46.1	47.78
64	γ	UMa ⬚	●	2.4	0.0	↓	0	**Phegda, Phad**	84	11 53.8	53.69
69	δ	UMa ⬚	●	3.3	0.1	↓	1	. . **Megrez** . .	81	12 15.4	57.03
77	ε	UMa ⬚	●	1.8	0.0	↓	0	. . . **Alioth** . . .	82	12 54.0	55.96
78		UMa ⬚	·	4.9	✩ 0.4	↓	3	. . . .	82	13 00.7	56.37
79	ζ	UMa ⬚	●	2.0	✩ 0.1	↓	0	Mizar ⎱ Sep.11′.8	80	13 23.9	54.93
80		UMa ⬚	·	4.0	0.2	↓	2	Alcor ⎰	80	13 25.2	54.99
85	η	UMa ⬚	●	1.9	−.1	↓	−1	**Alkaid, Benetnasch**	102	13 47.5	49.31

Mizar, Alcor: This binary is often called the horse and rider. Its 11.8 arc-minute separation is much greater than the limit of resolution of the eye with normal vision (approximately 5′). Therefore, Alcor should be well visible when the sky is dark enough. Other stars testing the resolution of the unaided eye are ϑ Tau (Chart E3), α Cap = Algiedi (E22), μ Sco (S21), and δ Gru (S24).

BINARY		Position	V-Mag.		B−V		Te.	Sep.	PA	Vis.
78		UMa ⬚	· 5.0	7.8	0.3	0.7	↓↓′0	1″.4	●●	⊙
							2015	1.1	●●	◯
79	ζ	UMa ⬚	● 2.3	3.9	0.0	0.1	↓↓	14.4	✦	⊙

VARIABLE STAR

VY UMa ⬚ · irregular
Extrema 5.8–7.0
Color orange-red.

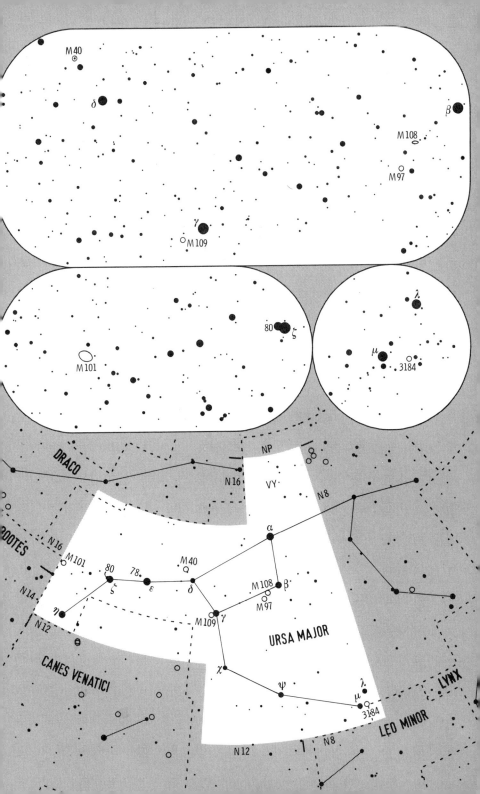

NEBULA	Position		v-Mag.	Size		Shape	Type	Vis.	Dist.	R.A.	Dec.
4244	CVn	♎	10½	13/□′	15′	| Sd	Glx	○	16 Mly	12ʰ17ᵐ.5	37°.81
4258 M 106	CVn	♎	8½	13	12	◖ Sc	Glx	⬮	30 M	12 19.0	47.30
4449	CVn	♀	9½	12	4	◯ Ir	Glx	⬩	16 M	12 28.2	44.09
4490	CVn	♀	10	12	5	◖ Sd	Glx	⬩	40 M	12 30.6	41.64
4736 M 94	CVn	♀	8½	11	4	◯ Sa	Glx	⬩	20 M	12 50.9	41.12
5055 M 63	CVn	♎	9	12	6	◖ Sc	Glx	⬩	30 M	13 15.8	42.04
5194 M 51	CVn	♎	8½	13	8	◯ Sc	Glx	⬮	30 M	13 29.9	47.20
5195	CVn	♎	10	12	3	◯ Ir	Glx	⬩	30 M	13 30.0	47.27

4244 ..... Very faint galaxy, but intriguing edge-on shape, elongated core.

4258 M 106 Elliptical glow in binoculars, a bright central region with stellar core in a telescope, faint dust features, traces of spiral arms; the mag. 11½ galaxies NGC 4217 and 4220 lie 35′ west and 45′ northwest.

4449 ..... Approximately rectangular, bright elongated central region, dust features barely visible; the halo is brightest at the northeast side.

4490 ..... Elongated central region within a large faint background glow; a light bridge extends to the mag. 12 galaxy NGC 4485 only 4′ north.

4736 M 94 Almost stellar in binoculars, bright round core in a telescope, faint halo elongated east-west, hints of spiral arms, medium power best.

5055 M 63 Distinct nonstellar core, oval halo, spiral arms not observable.

5194 M 51 **Whirlpool Galaxy**, bright core easily visible, two long spiral arms observable in a telescope, one arm is winding towards NGC 5195, wonderful galaxy, this might be the most beautiful spiral in the sky.

5195 ..... Probably companion galaxy of M 51, seems to touch M 51, the cores are only 5′ apart, looks hardly fainter but clearly smaller than M 51.

STAR	Position		V-Mag.	B–V	Te.	Abs.	Name	Dist.	R.A.	Dec.
53 ξ	UMa	⬛	• 3.8 ✶	0.6	⬇	4ᴹ	Alula Australis	25 ly	11ʰ18ᵐ.2	31°.53
54 ν	UMa	⬛	• 3.5	1.4	⬇	−2	Alula Borealis	420	11 18.5	33.09
8 β	CVn	♀	• 4.2	0.6	⬇	5	....	27.3	12 33.7	41.36
Y	CVn	♎	· 5.2–5.6	2.9	⬇	−2	. La Superba .	800	12 45.1	45.44
12 α	CVn	♀	• 2.8 ✶	−.1	⬇	0	. Cor Caroli .	110	12 56.0	38.32
17,15	CVn	♞	· 5.3 ✶	0.1	⬇	−1	.....	200, 1000	13 09.9	38.51
25	CVn	⬛	· 4.8 ✶	0.2	⬇	1	....	190	13 37.5	36.29

BINARY	Position	V-Mag.		B–V		Te.	Sep.	PA	Vis.
53 ξ	UMa ⬛	• 4.3	4.8	0.6	0.6	⬆⬆	’0	1″.8	•• ⬩

VARIABLE STAR	
Y	CVn ♎ · semireg.
Period	≈ 157 d
Extrema	4.9–6.0

	1995		2003	1.8	•• ⬩
2006	1.7	•. ⬩
2009	1.6	•. ⬩
2012	1.7	⦂ ⬩
2020 ←	2015	1.8	⦂ ⬩

It is a red giant of 400 million km = 250 million miles diameter; color is distinct only in a telescope.

12 α	ĊVn ♀	• 2.9	5.5	−.1	0.3	⬆⬆	19.3	•. ⬩
17,15	CVn ♞	· 5.9	6.3	0.3	−.1	⬆⬆	278	•· ⬔
25	CVn ⬛	· 5.0	7.0	0.2	0.3	⬆⬆	1.8	•• ⬩

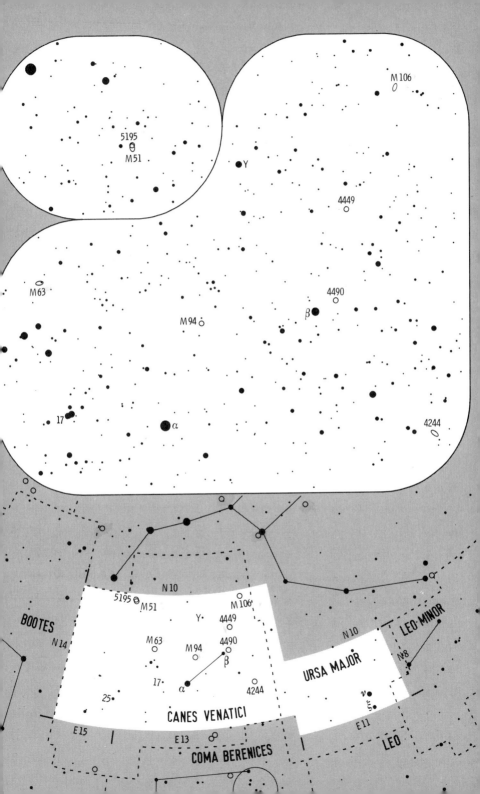

NEBULA		Position	v-Mag.	Size		Shape	Type	Vis.	Dist.	R.A.	Dec.
6205	M 13	Her	⊡	6	12/□′	15′	O V	GC	⊗	25 000 ly	16ʰ41ᵐ7	36°46
6341	M 92	Her	⊡	6½	11	8	O IV	GC	⊗	30 000	17 17.1	43.14

6205 M 13 **Hercules Cluster**, bright nebula in binoculars, outer portion is well resolved in a telescope at high power, core is partially resolved.
6341 M 92 Similar to M 13, some outer stars resolved in a telescope, oval halo.

STAR			Position	V-Mag.	B–V	Te.	Abs.	Name	Dist.	R.A.	Dec.
17	κ	Boo	⊡ •	4.4	✴ 0.2	↓	1ᴹ	} Sep. 36′ •●	165 ly	14ʰ13ᵐ5	51°79
21	ι	Boo	⊡ •	4.7	✴ 0.2	↓	2	}	98	14 16.2	51.37
23	ϑ	Boo	⊡ •	4.0	0.5	↓	3	. . . .	48	14 25.2	51.85
27	γ	Boo	⊡ ●	3.0	0.2	↓	1	. . Ceginus . .	86	14 32.1	38.31
39		Boo	⊡ ·	5.7	✴ 0.5	↓	1	. . . .	230	14 49.7	48.72
42	β	Boo	⊡ ●	3.5	1.0	↓	−1	. . Nekkar . .	220	15 01.9	40.39
44	i	Boo	⊡ ·	4.7–4.9✴	0.6	↓	4	. . . .	42	15 03.8	47.65
49	δ	Boo	⊡ ●	3.4	✴ 0.9	↓	1	. . . .	118	15 15.5	33.31
51	μ	Boo	⊡ •	4.2	✴ 0.3	↓	1	. Alkalurops .	120	15 24.5	37.38
7	ζ	CrB	⊡ •	4.6	✴ −.1	↓	−1	. . . .	450	15 39.4	36.64
11	φ	Her	⊡ •	4.2	−.1	↓	0	. . . .	230	16 08.8	44.93
17	σ	CrB	⊡ ·	5.2	✴ 0.6	↓	4	. . . .	71	16 14.7	33.86
22	τ	Her	⊡ •	3.9	−.1	↓	−1	. . . .	320	16 19.7	46.31
30 g		Her	⊡ ·	4.5–5.2	1.4	↓	−1	. . . .	360	16 28.6	41.88
35	σ	Her	⊡ •	4.2	0.0	↓	−1	. . . .	300	16 34.1	42.44
40	ζ	Her	⊡ ●	2.8	0.7	↓	3	. . . .	35	16 41.3	31.60
44	η	Her	⊡ ●	3.5	0.9	↓	1	. . . .	112	16 42.9	38.92
58	ε	Her	⊡ •	3.9	0.0	↓	0	. . . .	165	17 00.3	30.93
67	π	Her	⊡ ●	3.2	1.4	↓	−2	. . . .	360	17 15.0	36.81
68 u		Her	⊡ ·	4.8–5.5	−.1	↓	−2	. . . .	900	17 17.3	33.10
75	ϱ	Her	⊡ •	4.1	✴ 0.0	↓	−1	. . . .	400	17 23.7	37.15
85	ι	Her	⊡ •	3.8	−.2	↓	−2	. . . .	500	17 39.5	46.01
91	ϑ	Her	⊡ •	3.9	1.4	↓	−3	. . . .	600	17 56.3	37.25

BINARY			Position	V-Mag.		B–V		Te.	Sep.	PA	Vis.
17	κ	Boo	⊡ •	4.5	6.6	0.2	0.4	↿↾	13.6″	••	⊡
21	ι	Boo	⊡ •	4.8	8.1	0.2	0.8	↿↾	38.8	••	⊡
39		Boo	⊡ ·	6.2	6.6	0.5	0.5	↿↾	2.7	••	⊡
44	i	Boo	⊡ ·	5.1	6–7	0.6	0.7	↿↾ ’0	2.2	••	⊡
			↰ 1995				2007	2.3	••	⊡
			2020 ↴ •				2015	2.1	••	⊡
49	δ	Boo	⊡ ●	3.5	7.8	1.0	0.6	↿↾	104.9	•●	⊡
51	μ	Boo	⊡ ●	4.3	6.5✴	0.3	0.6	↿↾	108.8	!	⊡
				7.0	7.6	0.6	0.6	↿↾	2.2	⦂	⊡
7	ζ	CrB	⊡ ·	5.0	6.0	−.1	−.1	↿↾	6.3	•●	⊡
17	σ	CrB	⊡ ·	5.6	6.6	0.6	0.6	↿↾	7.1	••	⊡
75	ϱ	Her	⊡ •	4.5	5.5	0.0	0.0	↿↾	4.1	•●	⊡

VARIABLE STAR

44 i Boo ⊡ · ◠◠◠
Period 0.267819 d
Min. 2451200.18
Binary star mag.
5.1 and 6.0–6.6.

30 g Her ⊡ · semireg.
Period 70–90 d
Extrema 4.3–6.3

68 u Her ⊡ · ◠◠
Period 2.05107 d
Min. 2451200.9
Eclipse ≈ 10 hours

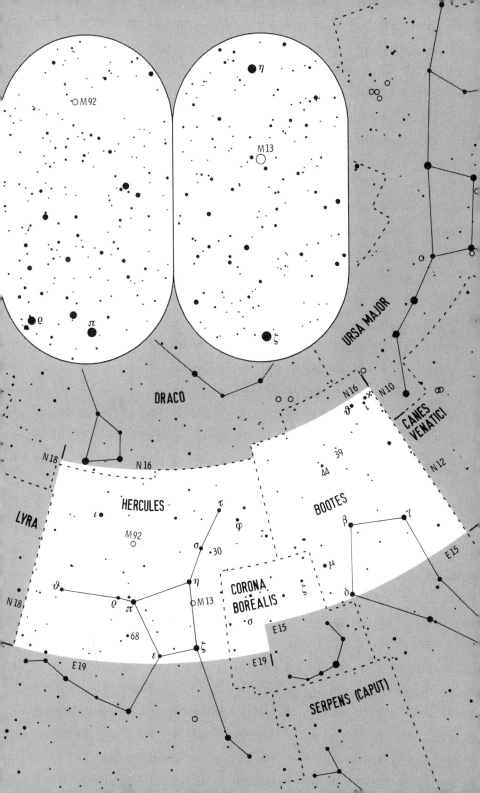

NEBULA	Position		v-Mag.	Size	Shape	Type	Vis.	Dist.	R.A.	Dec.
5866 M 102	Dra	⚲	10½ 11/□′	3′	⦿ S0	**Glx**	⦿	40 M ly	15ʰ06ᵐ.5	55°.76
5907	Dra	⚲	10½ 13	9	∣ Sc	**Glx**	⊙	40 M	15 15.9	56.33
6503	Dra	⚲	10½ 11	4	∥ Sd	**Glx**	⦿	16 M	17 49.5	70.14
6543	Dra	⚲	8 6	0.4	○ D	**PN**	⦿	3 000	17 58.6	66.63

5866 M 102 Appears as an elliptic nebula (see also comment at bottom right).
5907 ..... Difficult to see because of faintness, but distinct edge-on shape.
6503 ..... Elongated, nearly edge-on, northernmost nebula in this catalog.
6543 ..... Relatively easily visible, stellar in binoculars, blue-green oval in a
telescope at high power, 10′ northwest of the ecliptic's north pole.

STAR		Position		V-Mag.	B−V	Te.	Abs.	Name	Dist.	R.A.	Dec.
1	λ Dra	⦿	●	3.8	1.6	↟	−1ᴹ	.. Giauzar ..	330 ly	11ʰ31ᵐ.4	69°.33
5	κ Dra	⦿	●	3.9	−.1	↡	−2	....	520	12 33.5	69.79
11	α Dra	⦿	●	3.7	0.0	↡	−1	.. Thuban ..	310	14 04.4	64.38
12	ι Dra	⦿	●	3.3	1.2	↟	1	.. Edasich ..	102	15 24.9	58.97
13	ϑ Dra	⦿	●	4.0	0.5	↡	2	....	68	16 01.9	58.57
14	η Dra	⦿	●	2.7	0.9	↟	1	....	88	16 24.0	61.51
17,16	Dra	⦿	·	4.5 ✳	0.0	↡	−1	....	400	16 36.2	52.91
21	μ Dra	⦿	·	4.9 ✳	0.5	↟	3	....	88	17 05.3	54.47
22	ζ Dra	⦿	●	3.2	−.1	↡	−2	....	340	17 08.8	65.71
23	β Dra	⦿	●	2.8	1.0	↟	−3	. Rastaben .	370	17 30.4	52.30
25	ν Dra	⦿	·	4.1 ✳	0.3	↡	2	25 and 24 Dra	100	17 32.2	55.18
26	Dra	⦿	·	5.2 ✳	0.6	↟	4	....	46	17 35.0	61.87
31	ψ Dra	⚲	·	4.3 ✳	0.5	↡	3	....	72	17 41.9	72.15
32	ξ Dra	⦿	●	3.7	1.2	↟	1	. Grumium .	112	17 53.5	56.87
33	γ Dra	⦿	●	2.2	1.5	↟	−1	.. Ettanin ..	150	17 56.6	51.49
44	χ Dra	⚲	●	3.6	0.5	↡	4	....	26.3	18 21.1	72.73
39	Dra	⦿	·	4.9 ✳	0.1	↡	1	....	190	18 23.9	58.80
57	δ Dra	⦿	●	3.1	1.0	↟	1	.. Altais ..	100	19 12.6	67.66
63	ε Dra	⦿	●	3.8 ✳	0.9	↟	1	....	145	19 48.2	70.27

BINARY		Position		V-Mag.		B−V		Te.	Sep.	PA	Vis.
17,16	Dra	⬜	·	5.1✳ 5.5		0.0	−.1	⇅	90″.2	⦿	⦂
17	Dra			5.4 6.4		0.0	0.1	⇅	3.2	·⦁	⦿
21	μ Dra	⬜	·	5.6 5.7		0.5	0.5	⇊ ′0	2.2	⦿	⦿
							2015	2.4	⦿	⦿
25	ν Dra	⬜	●	4.9 4.9		0.3	0.3	⇅	61.9	●·	⦂
26	Dra	⬜	·	5.3 8.0		0.6	1.0	⇊ ′0	1.7	●·	⦿
				1995			2007	1.3	●·	⦿
				2020			2015	0.6	●·	○
31	ψ Dra	⚲	●	4.6 5.8		0.4	0.5	⇅	30.1	⦿	⦿
39	Dra	⬜	·	5.0✳ 7.9		0.1	0.5	⇅	88.9	⦿	⦿
				5.1 7.8		0.1	0.4	⇅	3.8	⦿	⦿
63	ε Dra	⦿	●	3.9 6.9		0.9	0.6	⇊	3.2	⦿	⦿

Comment on M 102

Messier's logbook contains a galaxy close to NGC 5866 as his entry number 102. Yet his description suggests a double observation of M 101. Did he make an error of 1ʰ in recording the right ascension? The designation M 102 is thus ambiguous.

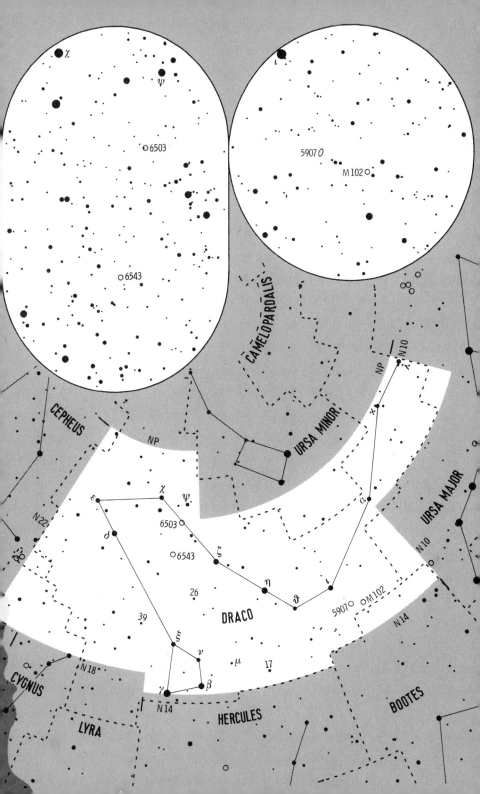

NEBULA	Position		v-Mag.	Size	Shape	Type	Vis.	Dist.	R.A.	Dec.
6720 M 57	Lyr	◙	8½	9/□′	1.5	O R	**PN**	⚬	1 800 ly	18ʰ53ᵐ6	33°03
6779 M 56	Lyr	◙	8½	12	5	O X	**GC**	⚬	30 000	19 16.6	30.18
6826	Cyg	◙	8½	7	0.5	O D	**PN**	⊙	3 000	19 44.8	50.52

6720 M 57 **Ring Nebula**, the most famous planetary nebula, easy to find, looks almost like a star in binoculars, shows a disk in a telescope at low power and a nice oval ring at high power, takes high power well, central region not black; a nebula filter improves the contrast.

6779 M 56 Dim globular cluster, hard to resolve into stars, rich background.

6826 ..... **Blinking Planetary**, stellar in binoculars, a disk in a telescope at high power; averted vision shows the nebula well, which disappears with direct vision as the mag. 10.4 central star becomes visible.

STAR		Position	V-Mag.	B−V	Te.	Abs.	Name	Dist.	R.A.	Dec.
1	κ Lyr	◙ •	4.3	1.2	↓	0ᴹ		240 ly	18ʰ19ᵐ9	36°06
3	α Lyr	◙ ●	0.0	0.0	↓	1	. . . **Vega** . . .	25.3	18 36.9	38.78
5,4	ε Lyr	◙ •	3.9	☆ 0.2	↓	0	Double Double	160	18 44.4	39.64
6,7	ζ Lyr	◙ •	4.1	☆ 0.2	↓	1	....	155	18 44.8	37.60
10	β Lyr	◙ •	3.3–4.2☆	0.0	↓	−4	. . **Sheliak** . .	800	18 50.1	33.36
12	δ² Lyr	◙ •	4.2	1.5	↓	−3	....	900	18 54.5	36.90
13	Lyr	◙ •	4.0–4.3	1.5	↓	−1	. . R Lyrae . .	350	18 55.3	43.95
14	γ Lyr	◙ ●	3.2	0.0	↓	−3	. Sulaphat .	700	18 58.9	32.69
1	κ Cyg	◙ ●	3.8	0.9	↓	1	....	122	19 17.1	53.37
10	ι Cyg	◙ ●	3.8	0.1	↓	1	....	123	19 29.7	51.73
6	β Cyg	◙ ●	2.9	☆ 0.9	↓	−3	. . **Albireo** . .	390	19 30.7	27.96
13	ϑ Cyg	◙ •	4.3	☆ 0.5	↓	0	 61,600		19 36.4	50.22
16	Cyg	◙ ·	5.4	☆ 0.6	↓	4	....	70	19 41.8	50.52
18	δ Cyg	◙ ●	2.9	☆ 0.0	↓	−1	....	170	19 45.0	45.13
	χ Cyg	◙ ·	6.0–12	1.9	↓	1	....	340	19 50.6	32.91
24	ψ Cyg	◙ ·	4.9	☆ 0.1	↓	0	....	290	19 55.6	52.44
21	η Cyg	◙ •	3.9	1.0	↓	1	....	140	19 56.3	35.08

BINARY		Position	V-Mag.		B−V		Te.	Sep.	PA	Vis.
5,4	ε Lyr	◙ •	4.6☆	4.7☆	0.2	0.2	↓↓	209″0	⦂	▦
5	ε² Lyr		5.2	5.5	0.2	0.2	↓↓	2.4	••	⊙
4	ε¹ Lyr		5.0	6.1	0.1	0.3	↓↓ ’0	2.5	⦂	⊙
							2015	2.4	⦂	⊙
6,7	ζ Lyr	◙ •	4.3	5.7	0.2	0.3	↓↓	43.7	⦂	⚬
10	β Lyr	◙ •	3–4	7.2	0.0	−.1	↓↓	45.7	⦂	⚬
6	β Cyg	◙ ●	3.1	5.1	1.1	−.1	↓↓	34.5	••	⚬
13	ϑ Cyg	◙ •	4.5	6.5	0.4	1.0	↓↓	300	••	▦
16	Cyg	◙ ·	6.0	6.2	0.6	0.7	↓↓	39.6	•⦂	⚬
18	δ Cyg	◙ •	2.9	6.5	0.0	0.3	↓↓ ’0	2.6	•⦁	⊙
							2015	2.7	•⦁	⊙
24	ψ Cyg	◙ ·	5.0	7.4	0.1	0.3	↓↓	3.0	⦂	⊙

VARIABLE STAR

10 β Lyr	◙ •	⌒⌒
Period		12.94 d
Min.		2451203
2nd min. mag.		3.8
13 R Lyr	◙ •	semireg.
Period		≈ 46 d
Extrema		3.9–5.0
χ Cyg	◙ ·	⟋‾⟍
Period		407 d
Max.		2451532
Min.	Max.	+240
Extrema		3.3–14.2

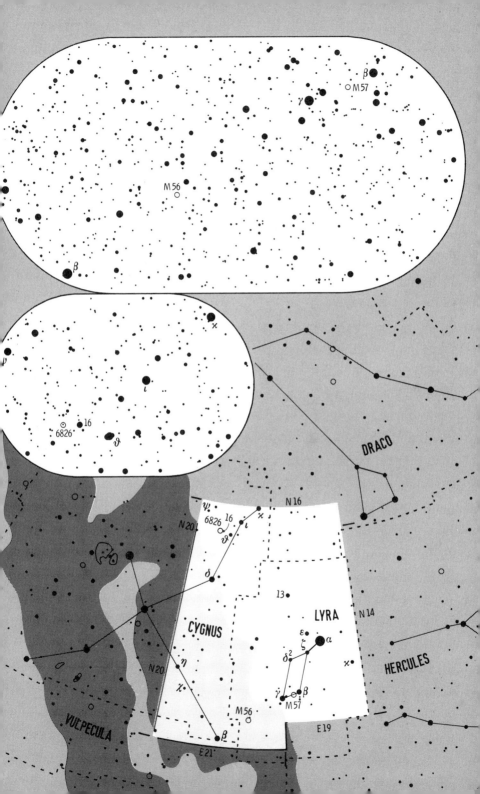

NEBULA	Position	v-Mag.	Size	Shape	Type	Vis.	Dist.	R.A.	Dec.
6913 M 29	Cyg	7	11/□' 6'	O p n	OC		4 000 ly	20ʰ23ᵐ.9	38°.53
6940	Vul	6½	13	25	O m	OC		2 500	20 34.6	28.30
6960	Cyg	9	14	60	| Fi	DN		1 500	20 45.7	30.72
IC 5067 ...	Cyg	7	14	60	O Em	DN		2 500	20 50.0	44.00
6992	Cyg	7½	14	60	❪ Fi	DN		1 500	20 57.0	31.40
7000	Cyg	5	14	120	O Em	DN		2 500	20 58.8	44.33
7027	Cyg	8½	5	0.3	O A	PN		3 500	21 07.0	42.24

6913 M 29 Contains only a few bright stars in a rich field, needs low power.

6940 ..... Large nebulous patch in binoculars, nicely resolved in a telescope.

6960 ..... **Veil Nebula, Cirrus Nebula, Filamentary Nebula**, dim filaments on both sides of the foreground star 52 Cygni, see NGC 6992.

IC 5067 ... **Pelican Nebula**, invisible except at lowest power, detail is only visible through a nebula filter, a tough test object for very dark sky.

6992 ..... **Veil Nebula, Cirrus Nebula, Network Nebula**, supernova remnant, slightly easier than NGC 6960, dark sky and low power essential, impressive filaments through a nebula filter (NGC 6992–6995).

7000 ..... **North America Nebula**, may be visible by unaided eye, almost too large for a telescope, nebula filter recommended, region of highest contrast is "Mexico"; the northern part merges into the Milky Way.

7027 ..... Relatively easily visible as a star, but only at high power as a disk.

STAR		Position	V-Mag.	B–V	Te.	Abs.	Name	Dist.	R.A.	Dec.
31	o¹	Cyg	• 3.4	✶ 0.7	↓	−5ᴹ	31, 30 Cyg	1 400, 750 ly	20ʰ13ᵐ.6	46°.76
29		Cyg	• 4.7	✶ 0.3	↓	−3	.....	135, 2 000	20 14.6	36.80
32	o²	Cyg	• 4.0	1.5	↓	−4	....	1 000	20 15.5	47.71
34		Cyg	• 4.7–4.9	0.4	↓	−7	. . P Cygni . .	5 000	20 17.8	38.03
37	γ	Cyg	● 2.2	0.7	↓	−6	. . . Sadr . . .	1 400	20 22.2	40.26
41		Cyg	• 4.0	0.4	↓	−3	....	750	20 29.4	30.37
46	ω²	Cyg	· 5.1	✶ 1.0	↓	−1	....	430	20 31.2	49.22
50	α	Cyg	● 1.3	0.1	↓	−8	. . **Deneb** . .	2 000	20 41.4	45.28
52		Cyg	• 4.2	1.0	↓	0	near NGC 6960	205	20 45.7	30.72
53	ε	Cyg	• 2.5	1.0	↓	1	....	72	20 46.2	33.97
54	λ	Cyg	• 4.5	✶ −.1	↓	−3	....	900	20 47.4	36.49
T		Vul	· 5.4–6.1	0.7	↓	−4	....	2 000	20 51.5	28.25
58	ν	Cyg	• 3.9	0.0	↓	−1	....	350	20 57.2	41.17
62	ξ	Cyg	• 3.7	1.6	↓	−4	....	1 200	21 04.9	43.93
64	ζ	Cyg	● 3.2	1.0	↓	0	....	150	21 12.9	30.23

BINARY		Position	V-Mag.		B–V		Te.	Sep.	PA	Vis.
31	o¹	Cyg	• 3.8	4.8	1.2	0.1	↓↓	336″	•	
			" 7.0		" −.1		↓	107.0	•	
29		Cyg	• 4.9	6.6	0.1	1.3	↓↓	216	•	
46	ω²	Cyg	• 5.4	6.6	1.6	0.0	↓↓	256.9	••	
54	λ	Cyg	• 4.8	6.1	−.2	0.2	↓↓	0.9	•	

VARIABLE STAR

34 P Cyg · irregular
Extrema 3.0–6.0

T Vul ·
Period 4.4355 d
Max. 2451201.5

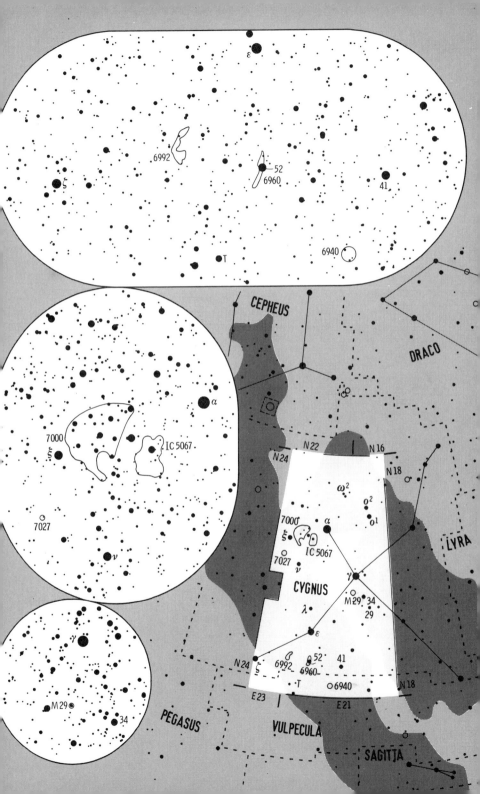

ε

6992

6960
52

41

6940

T

CEPHEUS

DRACO

7000

ξ

IC 5067

7027

ν

α

N 22   N 16

N 24   N 18

ω²

o²

o¹

7000
ξ
IC 5067
7027
ν

α

γ

LYRA

CYGNUS

λ

M 29 34
29

ε

N 24   52   41
ξ   6992
6960   N 18

E 23   E 21

T   6940

PEGASUS

γ

M 29

34

VULPECULA

SAGITTA

NEBULA	Position		v-Mag.	Size	Shape	Type	Vis.	Dist.	R.A.	Dec.
6939	Cep	⌷	8	13/□′	10′	O m	OC	⌾	5000 ly	20ʰ31ᵐ4	60°63
6946	Cep	⌷	9	14	10	O Sd	Glx	⌾	20 M	20 34.8	60.17
IC 1396 ...	Cep	⌷	4	12	50	O m n	OC	⌾	3000	21 39.1	57.50
7654 M 52	Cas	⌷	7	12	12	O r	OC	⌾	5000	23 24.2	61.58
7789	Cas	⌷	7	13	15	O r	OC	⌾	6000	23 57.0	56.73

6939	Faint open cluster, nebulous glow, very hard to resolve into stars.
6946	Difficult object, a galaxy without a central core or other features.
IC 1396 ...	Sparse, inconspicuous in a telescope, better in binoculars or a finder; surrouding diffuse nebula (dashed) visible through a nebula filter.
7654 M 52	Nebulous glow in binoculars, many faint stars resolved in a telescope.
7789	Tremendous number of stars for an open cluster, only resolved in a telescope, but the background still remains irregularly nebulous.

STAR	Position		V-Mag.	B–V	Te.	Abs.	Name	Dist.	R.A.	Dec.
3 η	Cep	⌷	● 3.4	0.9	↓	3ᴹ	. . . .	47 ly	20ʰ45ᵐ3	61°84
T	Cep	⌷	· 5.6–10	1.3	↓	−1	. . . .	700	21 09.5	68.49
5 α	Cep	⌷	● 2.5	0.3	↓	2	. Alderamin .	49	21 18.6	62.59
8 β	Cep	⌷	● 3.2	✳ −.2	↓	−3	. . Alfirk . .	600	21 28.7	70.56
μ	Cep	⌷	● 3.9–4.5	2.3	●	−6	. (giant star) .	3000	21 43.5	58.78
17 ξ	Cep	⌷	● 4.3	✳ 0.4	↓	2	. . . .	100	22 03.8	64.63
21 ζ	Cep	⌷	● 3.4	1.6	↓	−4	. . . .	800	22 10.9	58.20
23 ε	Cep	⌷	· 4.2	0.3	↓	2	. . . .	84	22 15.0	57.04
27 δ	Cep	⌷	● 3.4–4.2 ✳	0.5	↓	−4	. . . .	1000	22 29.2	58.42
32 ι	Cep	⌷	● 3.5	1.0	↓	1	. . . .	116	22 49.7	66.20
33 π	Cep	⌷	· 4.4	✳ 0.8	↓	0	. . . .	225	23 07.9	75.39
34 o	Cep	⌷	· 4.7	✳ 0.8	↓	1	. . . .	210	23 18.6	68.11
4	Cas	⌷	· 4.9	✳ 1.7	↓	−2	. . . .	750	23 24.8	62.28
AR	Cas	⌷	· 4.8	✳ −.1	↓	−2	. . . .	600	23 30.0	58.55
35 γ	Cep	⌷	● 3.2	1.0	↓	3	. . Errai . .	45	23 39.3	77.63
7 ϱ	Cas	⌷	· 4.4–4.6	1.2	↓	−8	. . . .	6000	23 54.4	57.50
8 σ	Cas	⌷	· 4.9	✳ −.1	↓	−4	. . . .	1500	23 59.0	55.76

BINARY	Position		V-Mag.		B–V	Te.	Sep.	PA	Vis.
8 β	Cep	⌷	● 3.2	7.9	−.2	0.1	↕	13″3	●• ⌷
17 ξ	Cep	⌷	· 4.4	6.5	0.3	0.5	↕′0	7.9	●• ⌷
							2015	8.0	●• ⌷
27 δ	Cep	⌷	● 4	6.3	0.6	0.0	↓↕	40.7	● ⌷
33 π	Cep	⌷	· 4.5	6.8	0.8	0.5	↕′0	1.1	● ⌷
							2015	1.1	● ⌷
34 o	Cep	⌷	· 4.9	7.1	0.8	0.5	↕′0	3.3	●• ⌷
							2015	3.4	●• ⌷
4	Cas	⌷	· 5.0	7.6	1.7	1.4	↕↕	95.9	●• ⌷
AR	Cas	⌷	· 4.9	7.0	−.1	0.0	↕↕	75.6	●• ⌷
8 σ	Cas	⌷	· 5.0	7.2	−.1	−.1	↕↕	3.2	●• ⌷

VARIABLE STAR

T	Cep	⌷ ⌒⌒
	Period	400 d
	Max.	2451354
	Extrema	5.2–11.2
μ	Cep	⌷ ● semireg.
	Extrema	3.4–5.1
27 δ	Cep	⌷ ● ⟋‾
	Period	5.3663 d
	Max.	2451203.0
7 ϱ	Cas	⌷ · semireg.
	Extrema	4.1–6.2

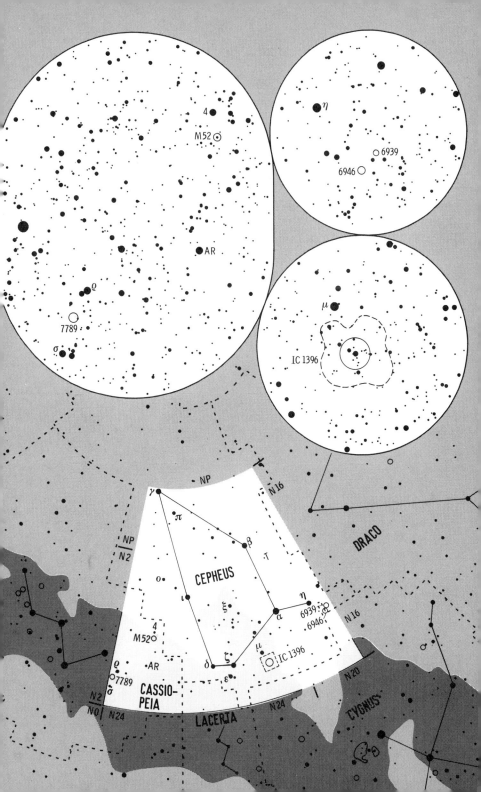

4
M52 ⊙

η

○ 6939
6946 ○

μ

IC 1396

7789

σ

ρ

AR

NP

γ

π

β

T

DRACO

o

ι

CEPHEUS

ξ

η
α
6939
6946
N 16

N 16

μ

IC 1396

4
M52 ○

ρ

AR

ζ

δ

ε

N 20

7789

σ

CASSIO-
PEIA

N 2
N 0   N 24

LACERTA

N 24

CYGNUS

σ

NEBULA		Position		v-Mag.	Size	Shape	Type	Vis.	Dist.	R.A.	Dec.
7092	M 39	Cyg	⊙	5	12/□′	30′	O p	OC	⦂⦂	1 000 ly	21ʰ32ᵐ.2	48°.43
7209	.....	Lac	⊙	7	13	20	O m	OC	⊙⊙	3 000	22 05.2	46.50
7243	.....	Lac	⊙	6½	13	20	O p	OC	⊙⊙	3 000	22 15.3	49.88
7662	.....	And	ρ	8½	7	0.5	O R	PN	⊙	4 000	23 25.9	42.54

7092 M 39 Consists of a few bright stars which are well resolved in binoculars or in a finder, rather disappointing view in a telescope, triangular.

7209 ..... Binoculars show few stars within a nebulous background, which resolves into a number of stars in a telescope, interesting patterns.

7243 ..... Partially resolved in binoculars, irregular shape; the main part consists of four groups of stars arranged nicely along a semicircle.

7662 ..... **Blue Snowball**, visible in binoculars as a star; a telescope at high power reveals a disk or a ring with a non-black center; the brightest section is at the northeast edge; the color is distinctly blue-green.

STAR		Position		V-Mag.	B−V	Te.	Abs.	Name	Dist.	R.A.	Dec.
61	Cyg	•	•	4.8 ⚹	1.1	↓	7ᴹ	Piazzi's Flying S.	11.4 ly	21ʰ06ᵐ.9	38°.75
65	τ Cyg	•	•	3.7	0.4	↓	2	....	68	21 14.8	38.05
67	σ Cyg	•	•	4.2	0.1	↓	−7	....	4 000	21 17.4	39.39
73	ρ Cyg	ρ	•	4.0	0.9	↓	1	} Sep. 25′ •	125	21 34.0	45.59
W	Cyg	ρ	·	5.4–6.2	1.5	↓	−1	....	600	21 36.0	45.37
79	Cyg	•	·	5.4 ⚹	0.0	↓	1	......	270,370	21 43.5	38.29
78	μ Cyg	•	•	4.4 ⚹	0.5	↓	2	......	73,250	21 44.2	28.75
81	π² Cyg	⊙	•	4.2	−.1	↓	−4	....	1 100	21 46.8	49.31
1	Lac	•	•	4.1	1.5	↓	−2	....	650	22 16.0	37.75
2	Lac	⊙	•	4.6	−.1	↓	−1	....	500	22 21.0	46.54
3	β Lac	•	•	4.4	1.0	↓	1	....	170	22 23.6	52.23
4	Lac	⊙	•	4.6	0.1	↓	−6	....	3 000	22 24.5	49.48
5	Lac	⊙	•	4.3	1.7	↓	−3	....	1 100	22 29.5	47.71
7	α Lac	⊙	•	3.8	0.0	↓	1	....	103	22 31.3	50.28
8	Lac	•	·	5.3 ⚹	−.2	↓	−2	....	1 000	22 35.9	39.63
1	o And	○	•	3.6	−.1	↓	−3	....	700	23 01.9	42.33
16	λ And	•	•	3.7–4.0	1.0	↓	2	....	83	23 37.6	46.46
17	ι And	ρ	•	4.3	−.1	↓	−2	....	480	23 38.1	43.27
19	κ And	ρ	•	4.1	−.1	↓	0	....	175	23 40.4	44.33

BINARY		Position		V-Mag.		B−V		Te.	Sep.	PA	Vis.
61	Cyg	•	·	5.2	6.0	1.1	1.3	↓↓	′0 31″.9	•	⊙
									2015 33.1	•	⊙
79	Cyg	•	·	5.7	7.0	0.0	0.1	↓↓	150	••	⦂⦂
78	μ Cyg	•	·	4.5⚹	6.9	0.5	0.4	↓↓	198	••	⊛
2020 ←·····→ 1995				4.8	6.2	0.5	0.6	↓↓	′0 1.8	•·	⊙
	•								2007 1.6	•·	⊙
									2015 1.3	•·	⊙
8	Lac	•	·	5.7	6.5	−.2	−.1	↓↓	22.4	•	⊙

VARIABLE STAR

W Cyg ⌷ semireg.
Period ≈ 130 d

16 λ And • semireg.
Period 54–56 d

Piazzi's Flying Star
Large proper motion of 5″.2/yr north-east.

7243

π²

M 39

α

4

5

2

7209

W ϱ

χ

ι

7662

ο

CASSIOPEIA

N2 N22

N22

N0

N20

β

λ

α 4

7243

π²

M 39

5

χ

2

7209

W ϱ

ι

7662

ο

LACERTA

CYGNUS

σ 61

ANDROMEDA

τ

N0

8

E 23

79

1

N20

μ

VULPECULA

PEGASUS

E 23

E0 ———— Equator, Ecliptic ———— Fall Constellations

NEBULA		Position	v-Mag.	Size		Shape	Type	Vis.	Dist.	R.A.	Dec.
247	.....	Cet	9	14/□'	18'	▯ Sd	Glx		10 Mly	$0^h47^m.1$	$-20°.76$
253	.....	Scl	7½	13	25	▯ Sc	Glx		10 M	0 47.6	−25.29
288	.....	Scl	8½	13	10	O X	GC		30000	0 52.8	−26.59
1068	M 77	Cet	9	11	3	O Sb	Glx		70 M	2 42.7	−0.01

247 ..... Large featureless galaxy, low power essential, a difficult object.
253 ..... **Sculptor Galaxy**, fantastic galaxy, elongated glow in binoculars, dust features become visible in a telescope; its core is small and oval.
288 ..... Hard object among globular clusters; a telescope resolves a few stars.
1068 M 77 Bright small Seyfert galaxy, active nucleus distinct at high power; the bright nucleus makes M 77 appear almost stellar in binoculars.

STAR		Position	V-Mag.	B−V	Te.	Abs.	Name	Dist.	R.A.	Dec.
κ^1	Scl	·	5.4 ✶	0.4	↓	1^M		220 ly	$0^h09^m.4$	$-27°.99$
8 ι	Cet	●	3.6	1.2	↓	−1		280	0 19.4	−8.82
T	Cet	·	5.3−6.1	1.7	↓	−2		750	0 21.8	−20.06
16 β	Cet	●	2.0	1.0	↓	0 Deneb Kaitos,		96	0 43.6	−17.99
α	Scl	·	4.3	−.1	↓	−2 . ⌊Diphda⌋ .		600	0 58.6	−29.36
31 η	Cet	●	3.5	1.2	↓	1		120	1 08.6	−10.18
37	Cet	·	5.0 ✶	0.5	↓	3		80	1 14.4	−7.92
45 ϑ	Cet	●	3.6	1.1	↓	1		115	1 24.0	−8.18
τ	Scl	·	5.7 ✶	0.3	↓	2		200	1 36.1	−29.91
52 τ	Cet	●	3.5	0.7	↓	6		11.9	1 44.1	−15.94
53 χ	Cet	·	4.5 ✶	0.4	↓	3		78	1 49.6	−10.69
55 ζ	Cet	●	3.7	1.1	↓	−1 Baten Kaitos		270	1 51.5	−10.33
59 υ	Cet	●	4.0	1.6	↓	−1		300	2 00.0	−21.08
66	Cet	·	5.5 ✶	0.6	↓	2		150	2 12.8	−2.39
68 o	Cet	●	3.4−9.2	1.4	↓	−2 ... **Mira** ...		400	2 19.3	−2.98
73 ξ²	Cet	·	4.3	−.1	↓	1		180	2 28.2	8.46
ω	For	·	4.9 ✶	0.0	↓	−1		440	2 33.8	−28.23
82 δ	Cet	●	4.1	−.2	↓	−3		700	2 39.5	0.33
86 γ	Cet	●	3.5 ✶	0.1	↓	1		82	2 43.3	3.24
89 π	Cet	●	4.2	−.1	↓	−1		450	2 44.1	−13.86
87 μ	Cet	●	4.3	0.3	↓	2		86	2 44.9	10.11
92 α	Cet	●	2.5	1.6	↓	−2 .. **Menkar** ..		220	3 02.3	4.09

BINARY		Position	V-Mag.		B−V		Te.	Sep.	PA	Vis.
κ^1	Scl	·	6.1	6.2	0.4	0.4	↿⇂	1".5	••	[·]
37	Cet	·	5.1	7.8	0.5	0.8	↿⇂	49.1	•'	[.']
τ	Scl	·	6.0	7.2	0.3	0.5	↿⇂'0	0.8	•:	[·]
						2015		0.8	•:	[·]
53 χ	Cet	·	4.7	6.7	0.3	0.6	↿⇂	184.0	••	[::]
66	Cet	·	5.7	7.6	0.6	0.7	↿⇂	16.7	••	[·]
ω	For	·	5.0	7.7	−.1	0.2	↿⇂	10.8	••	[°]
86 γ	Cet	●	3.5	7.0	0.1	0.5	↿⇂	2.7	••	[·]

VARIABLE STAR

T Cet [·] · semireg.
Period 159 d
Extrema 5.0−6.9
68 o Cet [·] • ⌢
Period 332 d
Max. 2451490
Min. Max.+205
Extrema 2.0−10.1

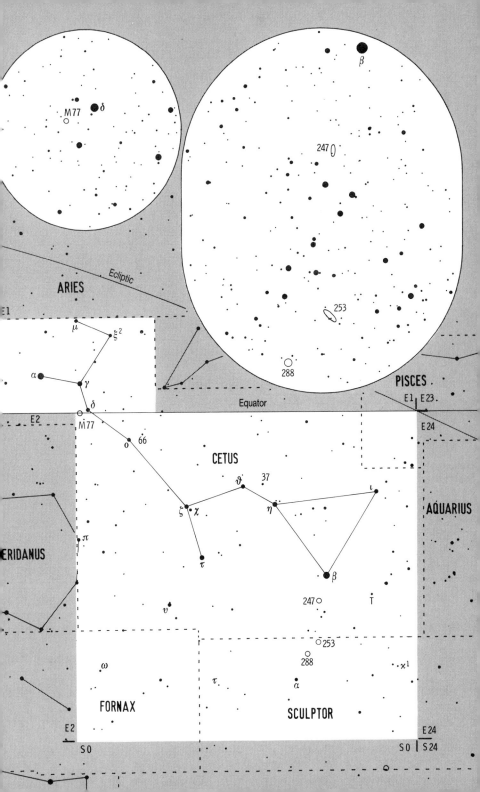

ARIES

M77

δ

Ecliptic

β

247

253

288

PISCES

E1   E23

E24

Equator

E1

E2

μ

ξ²

α

γ

δ

M77

o   66

CETUS

ϑ

37

χ

ζ

η

ι

AQUARIUS

τ

β

π

ERIDANUS

υ

247

T

253

288

ω

κ¹

τ

α

FORNAX

SCULPTOR

E2

E24

S0

S0 | S24

NEBULA	Position	v-Mag.	Size	Shape	Type	Vis.	Dist.	R.A.	Dec.
628 M 74	Psc ♂	9½ 14/□'	8'	O Sc	**Glx**	⊙°	50 Mly	1ʰ36ᵐ7	15°79

628 M 74 Very difficult except under darkest sky, lowest power essential, moderately bright core not exactly centered, see comment at the bottom.

STAR			Position	V-Mag.	B–V	Te.	Abs.	Name	Dist.	R.A.	Dec.
88	γ	Peg	• ●	2.8	−.2	↓	−2ᴹ	.. **Algenib** ..	360 ly	0ʰ13ᵐ2	15°18
35		Psc	• ·	5.8 ☆	0.3	↓	1	....	260	0 15.0	8.82
47		Psc	• •4.7–5.3		1.6	↓	−1	TV Piscium	500	0 28.0	17.89
34	ζ	And	• ●	4.1	1.1	↓	0	....	185	0 47.3	24.27
63	δ	Psc	• ●	4.4	1.5	↓	0	....	310	0 48.7	7.59
65		Psc	• ●	5.5 ☆	0.4	↓	0	....	350	0 49.9	27.71
36		And	• ·	5.5 ☆	1.0	↓	2	....	130	0 55.0	23.63
71	ε	Psc	• ·	4.3	1.0	↓	0	....	190	1 02.9	7.89
74	ψ¹	Psc	• ·	4.7 ☆	0.0	↓	0	....	240	1 05.7	21.47
86	ζ	Psc	• ·	4.9 ☆	0.4	↓	2	....	150	1 13.7	7.58
99	η	Psc	♂ ●	3.6	1.0	↓	−1	....	300	1 31.5	15.35
106	ν	Psc	• ●	4.4	1.4	↓	−1	....	370	1 41.4	5.49
110	o	Psc	• •	4.3	0.9	↓	0	....	250	1 45.4	9.16
1		Ari	• ·	5.8 ☆	0.7	↓	0	....	500	1 50.1	22.28
5	γ	Ari	• ●	3.9 ☆	0.0	↓	0	. Mesarthim .	200	1 53.5	19.29
111	ξ	Psc	• ·	4.6	0.9	↓	1	....	190	1 53.6	3.19
6	β	Ari	• ●	2.6	0.2	↓	1	. Sheratan .	59	1 54.6	20.81
9	λ	Ari	• ·	4.7 ☆	0.3	↓	2	....	134	1 57.9	23.60
113	α	Psc	• ●	3.8 ☆	0.1	↓	1	....	140	2 02.0	2.76
10		Ari	• ·	5.6 ☆	0.5	↓	2	....	170	2 03.7	25.94
13	α	Ari	• ●	2.0	1.1	↓	0	.. **Hamal** ..	66	2 07.2	23.46
41		Ari	• ●	3.6	−.1	↓	0	....	160	2 50.0	27.26
48	ε	Ari	• •	4.6 ☆	0.0	↓	0	....	300	2 59.2	21.34

BINARY			Position	V-Mag.		B–V		Te.	Sep.	PA	Vis.
35		Psc	•	·	6.0	7.6	0.3	0.4	⇊	11.″5	⦿
65		Psc	•	·	6.3	6.3	0.4	0.4	⇊	4.3	⦿
36		And	•	·	6.0	6.4	0.9	1.2	⇊'0	0.9	⦿
					2020				2007	1.0	⦿
					1995				2015	1.1	⦿
74	ψ¹	Psc	•	·	5.3	5.6	0.0	0.0	‖	29.8	⦿
86	ζ	Psc	•	·	5.2	6.3	0.3	0.5	‖	22.8	⦿
1		Ari	•	·	6.2	7.2	1.1	0.2	⇊	2.9	⦿
5	γ	Ari	•	•	4.6	4.7	0.0	0.0	‖	7.5	⦿
9	λ	Ari	•	•	4.8	7.3	0.3	0.6	‖	37.5	⦿
113	α	Psc	•	•	4.2	5.2	0.0	0.2	‖	1.8	⦿
10		Ari	•	·	5.8	7.7	0.5	0.7	⇊'0	1.2	⦿
									2015	1.4	⦿
48	ε	Ari	•	•	5.2	5.5	0.0	0.1	‖	1.5	⦿

VARIABLE STAR

47 TV Psc • · semireg.
Period 50–85 d

MESSIER Marathon
It is possible to observe all 110 Messier objects during just one night in March. M 74 is then the most difficult object at dusk, and M 30 the most difficult at dawn. *Advice:* Better take your time!

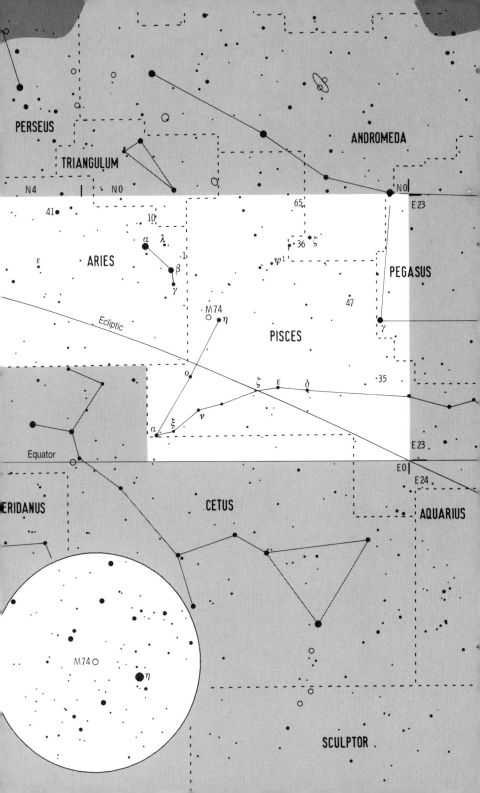

NEBULA Position v-Mag. Size Shape Type Vis. Dist. R.A. Dec.

			v-Mag	Size	Shape	Type	Vis	Dist	R.A.	Dec.
1360 ..	For	[symbol]	9	12/□′ 7′	0	D	PN	[•]	1500ly	$3^h33^m.3$ −25°.87
1535 ..	Eri	[symbol]	9½	6 0.3	0	D	PN	[•]	5000	4 14.3 −12.74

1360 .. Large planetary, true size 3 light years, less known among observers, near the limit of binoculars; a telescope shows mag. 11.3 central star, an extremely hot star with a surface temperature of about 100000 K.

1535 .. Bright central disk within a fainter oval halo, requires a telescope at very high power, the central star of magnitude 12.2 is difficult to see.

STAR Position V-Mag. B−V Te. Abs. Name Dist. R.A. Dec.

#		Pos	•	V-Mag	B−V	Te	Abs	Name	Dist	R.A.	Dec.
1	τ^1 Eri		•	4.5	0.5	↓	4^M	....	46 ly	$2^h45^m.1$	−18°.57
	β For		•	4.4	1.0	↓	1	....	170	2 49.1	−32.41
3	η Eri		•	3.9	1.1	↓	1	....	133	2 56.4	−8.90
11	τ^3 Eri		•	4.1	0.2	↓	2	....	86	3 02.4	−23.62
	α For		•	3.9 ✲	0.5	↓	3	....	46	3 12.1	−28.99
16	τ^4 Eri		•	3.7	1.6	↓	−1	....	260	3 19.5	−21.76
18	ε Eri		•	3.7	0.9	↓	6	....	10.5	3 32.9	−9.46
19	τ^5 Eri		•	4.3	−.1	↓	−1	....	310	3 33.8	−21.63
23	δ Eri		•	3.5	0.9	↓	4	....	29.5	3 43.2	−9.77
27	τ^6 Eri		•	4.2	0.4	↓	3	....	58	3 46.8	−23.25
32	Eri		•	4.5 ✲	0.7	↓	−1	....	350	3 54.3	−2.95
34	γ Eri		●	3.0	1.6	↓	−1	.. Zaurak ..	220	3 58.0	−13.51
38	o^1 Eri		•	4.0	0.3	↓	1	....	125	4 11.9	−6.84
39	Eri		·	4.9 ✲	1.2	↓	1	....	210	4 14.4	−10.26
40	o^2 Eri		•	4.4 ✲	0.8	↓	6	. (see below) .	16.5	4 15.3	−7.64
41	v^4 Eri		•	3.6	−.1	↓	0	....	180	4 17.9	−33.80
43	v^3 Eri		•	4.0	1.5	↓	−1	....	270	4 24.0	−34.02
52	v^2 Eri		•	3.8	1.0	↓	0	....	200	4 35.6	−30.56
48	ν Eri		•	3.9	−.2	↓	−2	....	600	4 36.3	−3.35
53	Eri		•	3.9	1.1	↓	1	....	108	4 38.2	−14.30
57	μ Eri		•	4.0	−.1	↓	−2	....	500	4 45.5	−3.25
67	β Eri		●	2.8	0.2	↓	1	.. Cursa ..	89	5 07.9	−5.09
	ε Col		•	3.9	1.1	↓	−1	....	270	5 31.2	−35.47
	α Col		●	2.6	−.1	↓	−2	.. Phact ..	260	5 39.6	−34.07
	β Col		●	3.1	1.2	↓	1	....	86	5 51.0	−35.77
	γ Col		•	4.4	−.2	↓	−3	....	850	5 57.5	−35.28
	δ Col		•	3.9	0.9	↓	0	....	235	6 22.1	−33.44

BINARY Position V-Mag. B−V Te. Sep. PA Vis.

		Pos	•	V-Mag	B−V	Te	Sep	PA	Vis
α For			•	4.0 6.9	0.5 0.8	↓↓	′0	4″9	•* [•]
				↗ 2020				2007 5.1	•* [•]
				• 1995				2015 5.4	•* [•]
32	Eri		•	4.8 6.1	0.9 0.1	↓↓		6.9	: [•]
39	Eri		•	4.9 8.0	1.2 0.7	↓↓		6.4	.• [•]
40	o^2 Eri		•	4.4 9.5	0.8 0.0	↓↓		83.4	•• [•]

Comment on o^2 Eri Its companion is the most easily observable white dwarf and the smallest star in this catalog; diameter only 20000km, 12000miles.

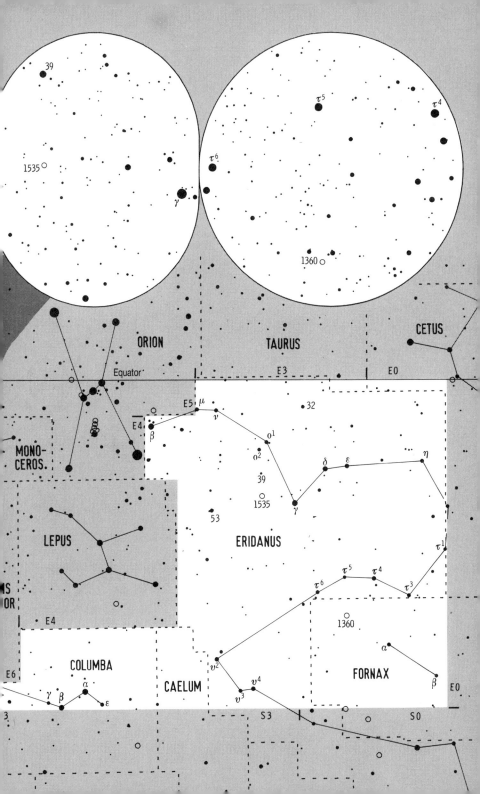

NEBULA	Position	v-Mag.	Size	Shape	Type	Vis.	Dist.	R.A.	Dec.
M 45	Tau 🔯	1½ 11/□′ 100′		O r n	**OC**	⚃	390 ly	3ʰ47ᵐ0	24°.12
Hyades ...	Tau 🔯	1 11	300	O m	**OC**	⚃	150	4 28	16.5
1647	Tau ♂	6½ 14	40	O m	**OC**	🎴	2 000	4 46.0	19.07
1952 M 1	Tau 🔯	8 11	6	0 Fi	**DN**	🎲	4 000	5 34.5	22.02

M 45   **Pleiades, Seven Sisters,** marvelous with unaided eye or binoculars, Merope's reflection nebula NGC 1435 visible under darkest sky.

Hyades ...   Only impressive with unaided eye or opera glasses, scattered stars, the closest and brightest star cluster, Aldebaran is a foreground star.

1647 .....   Large open cluster; it is resolved into many stars in binoculars.

1952 M 1   **Crab Nebula,** difficult in binoculars, elongated, irregular in a telescope, a nebula filter helps, the remnant of the supernova in 1054.

STAR		Position	V-Mag.	B−V	Te.	Abs.	Name	Dist.	R.A.	Dec.
1	o	Tau	• 3.6	0.9	⬇	−1ᴹ	} Sep. 55′	220 ly	3ʰ24ᵐ8	9°.03
2	ξ	Tau	• 3.7	−.1	⬇	0		220	3 27.2	9.73
17		Tau	• 3.7	−.1	⬇	−2	Electra ⎫	390	3 44.9	24.11
19		Tau	• 4.3	−.1	⬇	−1	Taygeta ⎪	390	3 45.2	24.47
20		Tau	• 3.8	−.1	⬇	−2	Maia ⎪ in	390	3 45.8	24.37
23		Tau	• 4.1	−.1	⬇	−1	Merope ⎬ M 45	390	3 46.3	23.95
25	η	Tau	● 2.8	−.1	⬇	−3	Alcyone ⎪	390	3 47.5	24.11
27		Tau	• 3.6	−.1	⬇	−2	Atlas ⎪	390	3 49.2	24.06
28	BU Tau		• 4.9–5.2	−.1	⬇	−1	Pleione ⎭	390	3 49.2	24.14
35	λ	Tau	● 3.4–3.9	−.1	⬇	−2	....	360	4 00.7	12.49
38	ν	Tau	• 3.9	0.0	⬆	1	....	132	4 03.2	5.99
47		Tau	· 4.8 ⚹	0.8	⬇	0	....	350	4 13.9	9.26
54	γ	Tau	• 3.6	1.0	⬆	0	⎫	155	4 19.8	15.63
61	δ¹	Tau	• 3.8	1.0	⬆	0	⎪	155	4 22.9	17.54
68	δ³	Tau	• 4.3 ⚹	0.0	⬇	1	⎬ in Hyades	150	4 25.5	17.93
74	ε	Tau	• 3.5	1.0	⬆	0	⎪	155	4 28.6	19.18
77	ϑ¹	Tau	• 3.8	1.0	⬆	0	⎪ } Sep. 5′.7	155	4 28.6	15.96
78	ϑ²	Tau	• 3.4	0.2	⬇	0	⎭	155	4 28.7	15.87
88		Tau	· 4.2 ⚹	0.2	⬇	1	....	150	4 35.7	10.16
87	α	Tau	● 0.9	1.5	⬆	−1	. **Aldebaran** .	66	4 35.9	16.51
94	τ	Tau	• 4.2 ⚹	−.1	⬇	−1	....	400	4 42.2	22.96
112	β	Tau	● 1.7	−.1	⬇	−1	**Elnath, Nath**	130	5 26.3	28.61
118		Tau	· 5.5 ⚹	0.0	⬇	−1	....	500	5 29.3	25.15
123	ζ	Tau	• 3.0	−.2	⬇	−3	....	400	5 37.6	21.14

BINARY		Position	V-Mag.	B−V	Te.	Sep.	PA	Vis.
47		Tau	· 4.9 7.3	0.8 0.8	⬆⬆	1″.3	⦂	⊡
68	δ³	Tau	· 4.4 7.6	0.0 0.6	⬇⬆	1.5	⦂	⊡
88		Tau	· 4.3 7.8	0.2 0.5	⬆⬆	69.6	••	⊡
94	τ	Tau	· 4.3 7.1	−.1 0.1	⬇⬇	62.9	•.	⊡
118		Tau	· 5.9 6.7	−.1 0.1	⬇⬇	4.7	•.	⊡

VARIABLE STAR

28 BU Tau 🔯 · irregular
35 λ Tau • ⎰‿⎱
Period 3.95295 d
Min. 2451201.4
2ⁿᵈ min. mag. 3.6

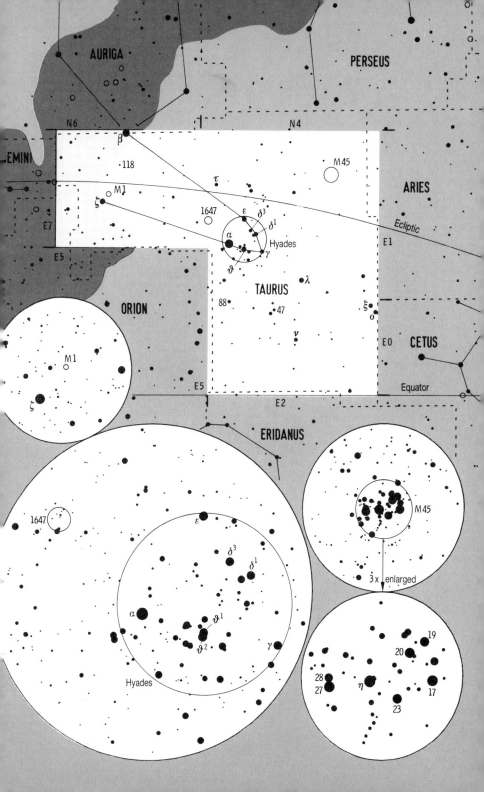

NEBULA	Position		v-Mag.	Size	Shape	Type	Vis.	Dist.	R.A.	Dec.
1904 M 79	Lep	⊙	8	12/□′	6′	O V	GC	✷	40 000 ly	5ʰ24ᵐ2 −24°53
1981	Ori	◰	4½	11	25	0 p n	OC	▩	1 400	5 35.2 −4.43
1973	Ori	◰	8	14	20	0 Re	DN	◎	1 400	5 35.3 −4.80
1976 M 42	Ori	◰	3½	11	40	0 Em	DN	⦂⦂	1 400	5 35.6 −5.43
1982 M 43	Ori	◰	8	13	12	0 Em	DN	⦂	1 400	5 35.6 −5.27

1904 M 79 Very difficult to resolve, well concentrated, far outside our galaxy.
1981 ..... Only a few bright stars, which are hard to recognize as a cluster.
1973 ..... Dim difficult object, the three sections are NGC 1973, 1975, 1977.
1976 M 42 **Orion Nebula**, primary nebula of all diffuse nebulae, impressive in every scope; dust clouds, bright arcs, and embedded stars are fantastic, color blue-green, contains famous trapezium; to an experienced observer a telescope can show more detail than many photographs.
1982 M 43 The northern part of the Orion Nebula, separated by a dust cloud.

STAR		Position	V-Mag.	B−V	Te.	Abs.	Name	Dist.	R.A.	Dec.
R	Lep	•	· 6.0–9.7	3.4	•	−2ᴹ	. . . .	1 000 ly	4ʰ59ᵐ6 −14°81
2 ε	Lep	•	● 3.2	1.5	↓	−1	. . . .	220	5 05.5 −22.37
RX	Lep	•	· 5.2–6.0	1.4	↓	−1	. . . .	450	5 11.4 −11.85
5 μ	Lep	•	● 3.2–3.4	−.1	↓	−1	. . . .	185	5 12.9 −16.21
4 κ	Lep	•	• 4.4	⚹ −.1	↓	−2	. . . .	550	5 13.2 −12.94
19 β	Ori	•	● 0.1	⚹ 0.0	↓	−7	. . **Rigel** . .	800	5 14.5 −8.20
20 τ	Ori	•	● 3.6	−.1	↓	−2	. . . .	500	5 17.6 −6.84
9 β	Lep	⊙	● 2.8	0.8	↓	−1	. . **Nihal** . .	160	5 28.2 −20.76
11 α	Lep	•	● 2.6	0.2	↓	−6	. . **Arneb** . .	1 400	5 32.7 −17.82
41 ϑ	Ori	◰	• 4.0	⚹ 0.0	↓	−4	41 and 43 Ori	1 400	5 35.3 −5.40
42,45	Ori	◰	• 4.1	⚹ 0.0	↓	−4	in NGC 1973	1 400, 380	5 35.4 −4.84
44 ι	Ori	◰	● 2.8	⚹ −.2	↓	−6	. . . .	1 400	5 35.4 −5.91
13 γ	Lep	•	• 3.5	⚹ 0.5	↓	4	. . . .	29.2	5 44.5 −22.45
14 ζ	Lep	•	• 3.6	0.1	↓	2	. . . .	70	5 47.0 −14.82
53 κ	Ori	•	● 2.1	−.2	↓	−5	. . **Saiph** . .	750	5 47.8 −9.67
15 δ	Lep	•	• 3.8	1.0	↓	1	. . . .	113	5 51.3 −20.88
16 η	Lep	•	• 3.7	0.3	↓	3	. . . .	49	5 56.4 −14.17

BINARY		Position	V-Mag.		B−V		Te.	Sep.	PA	Vis.
4 κ	Lep	•	• 4.4	7.1	−.1	0.3	‖	2″1	⦂	⊙
19 β	Ori	•	● 0.1	6.8	0.0	0.0	‖	9.5	⦂•	⊙
41 ϑ	Ori	◰	• 4.6⚹	4.8⚹	0.0	−.1	‖	140.0	•◦	▩
41 ϑ¹	Ori		5.1	6.7	0.0	0.1	‖	13.4	•◦	⊙
Trapezium			”	6.7	”	0.0	‖	12.7	•◦	⊙
			8.0	”	0.2	”	↓	8.7	◦•	⊙
43 ϑ²	Ori		5.1	6.4	−.1	−.1	‖	52.4	•◦	⦂⦂
42,45	Ori	◰	• 4.6	5.2	−.2	0.3	‖	252	•◦	▩
44 ι	Ori	◰	● 2.8	6.9	−.2	−.1	‖	11.3	•◦	⊙
13 γ	Lep	•	• 3.6	3.6	0.5	0.9	↓↓	97.2	⦂	⦂⦂

VARIABLE STAR

R Lep ▱ · ⌒⌒
Period ≈ 435 d
Max. ≈ 2451220
Min. Max. +200
Extrema 5.5–11.7
The reddest star.
RX Lep ▱ · irregular
Period 60–90 d
Extrema 5.0–7.4
5 μ Lep ▱ • irregular

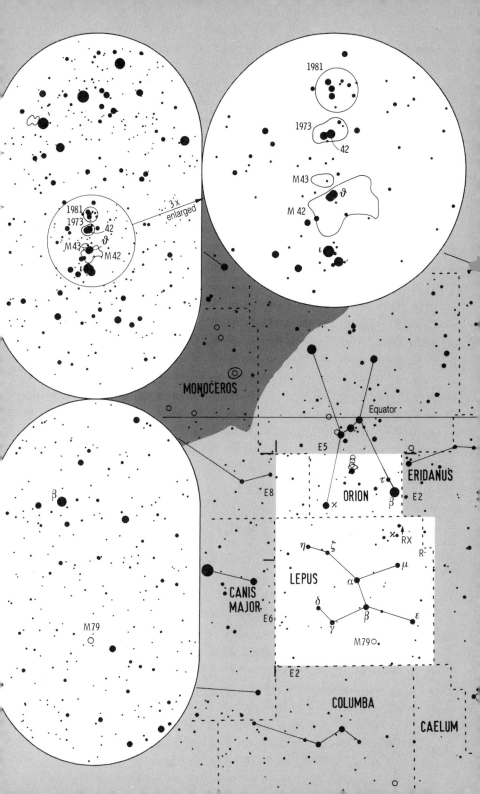

1981

1973

42

M43

M42

ϑ

ι

3x
enlarged

1981

1973

42

M43

ϑ

M 42

ι

MONOCEROS

Equator

E5

E8

ORION

τ

β

ERIDANUS

E2

χ

β

η

ζ

χ

RX

R

LEPUS

μ

α

CANIS
MAJOR

δ

β

ε

E6

γ

M79

M79

E2

COLUMBA

CAELUM

NEBULA	Position	v-Mag.	Size	Shape	Type	Vis.	Dist.	R.A.	Dec.
1788	Ori ☽	9	12/□′	6′	0 Re **DN**	☉	1400 ly	5ʰ06ᵐ9	−3°37
2024	Ori ♂	7½	13	20	0 Em **DN**	◎	1200	5 41.4	−1.87
2068 M78	Ori ♂	8	12	7	0 Re **DN**	☉	1200	5 46.7	0.07

1788 ..... A rare reflection nebula, very hard object, embedded mag. 10.1 star.
2024 ..... Relatively bright and rich in features, but Alnitak outshines it, needs clean optics, nebula filter helps, best if Alnitak outside field of view.
2068 M78 Brightest reflection nebula, appears like a comet, dark dust features, two embedded stars; 15′ north is mag. 10 reflection nebula NGC 2071.

STAR		Position	V-Mag.		B−V	Te.	Abs.	Name	Dist.	R.A.	Dec.
1	π³ Ori	•	●	3.2	0.5	↓	4ᴹ		26.2 ly	4ʰ49ᵐ8	6°96
3	π⁴ Ori	•	●	3.7	−.2	↓	−4		1100	4 51.2	5.61
8	π⁵ Ori	•	●	3.7	−.2	↓	−4		1100	4 54.3	2.44
14	Ori	•	·	5.3	☼ 0.3	↓	2		180	5 07.9	8.50
22	Ori	♂	•	4.4	☼ −.2	↓	−4		1200	5 21.7	−0.39
23	Ori	•	·	4.9	☼ −.1	↓	−3		1200	5 22.8	3.55
28	η Ori	♂	●	3.3–3.6	☼ −.2	↓	−5		1200	5 24.5	−2.40
24	γ Ori	•	●	1.6	−.2	↓	−3 .	Bellatrix	. 240	5 25.1	6.35
32	Ori	•	•	4.2	☼ −.1	↓	−1		300	5 30.8	5.95
33	Ori	•	·	5.4	☼ −.2	↓	−3		1200	5 31.2	3.29
34	δ Ori	♂	●	2.2	☼ −.2	↓	−6 .	. Mintaka . .	1200	5 32.0	−0.30
VV	Ori	♂	·	5.3–5.7	−.2	↓	−3		1200	5 33.5	−1.16
39	λ Ori	•	●	3.4	☼ −.2	↓	−5		1200	5 35.1	9.93
46	ε Ori	♂	●	1.7	−.2	↓	−6 .	. Alnilam . .	1200	5 36.2	−1.20
48	σ Ori	♂	•	3.6	☼ −.2	↓	−4		1200	5 38.8	−2.60
50	ζ Ori	♂	●	1.7	☼ −.2	↓	−6 .	. Alnitak . .	1200	5 40.8	−1.94
52	Ori	•	·	5.3	☼ 0.2	↓	0		450	5 48.0	6.45
58	α Ori	•	●0.3–0.9		1.8	↓	−5 .	Betelgeuse .	350	5 55.2	7.41

BINARY		Position	V-Mag.		B−V		Te.	Sep.	PA	Vis.
14	Ori	•	· 5.8	6.6	0.3	0.3	↓↓ ′0	0.8	•′	•
			1995 ⤳ 2020				2007	0.9	•′	•
			●				2015	1.0	•′	•
22	Ori	♂	• 4.7	5.7	−.2	−.1	↓↓ 241.9		•.	�
23	Ori	•	· 5.0	7.2	−.1	−.1	↓↓ 32.0		·•	☉
28	η Ori	♂	● 4	4.9	−.2	−.2	↓↓ 1.7		••	•
32	Ori	•	• 4.5	5.8	−.1	−.1	↓↓ 1.2		·•	•
33	Ori	•	· 5.8	6.9	−.2	−.1	↓↓ 1.9		·•	•
34	δ Ori	♂	● 2.2	6.8	−.2	−.2	↓↓ 52.4		⁞	•·
39	λ Ori	•	● 3.6	5.5	−.2	−.2	↓↓ 4.3		·•	•
48	σ Ori	♂	● 3.8	6.6	−.2	−.2	↓↓ 41.5		·•	☉
			"	6.6	"	−.2	↓ 12.9		·•	☉
50	ζ Ori	♂	● 1.9	4.0	−.2	−.2	↓↓ 2.4		⁞	•
52	Ori	•	· 6.0	6.0	0.1	0.3	↓↓ 1.0		•.	•

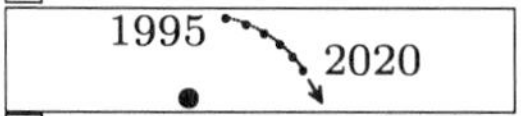

VARIABLE STAR

28 η Ori ♂ • ⟦‾_/‾⟧
Period 7.98928 d
Min. 2451202.27
Eclipse 15 hours
Binary star mag.
3.6–5.0 and 4.9.

VV Ori ♂ · ⟦‾_/‾⟧
Period 1.485376 d
Min. 2451200.51
Eclipse 6 hours

58 α Ori • ● semireg.
Periods 420 d
and ≈ 6 years
Extrema 0.0–1.3

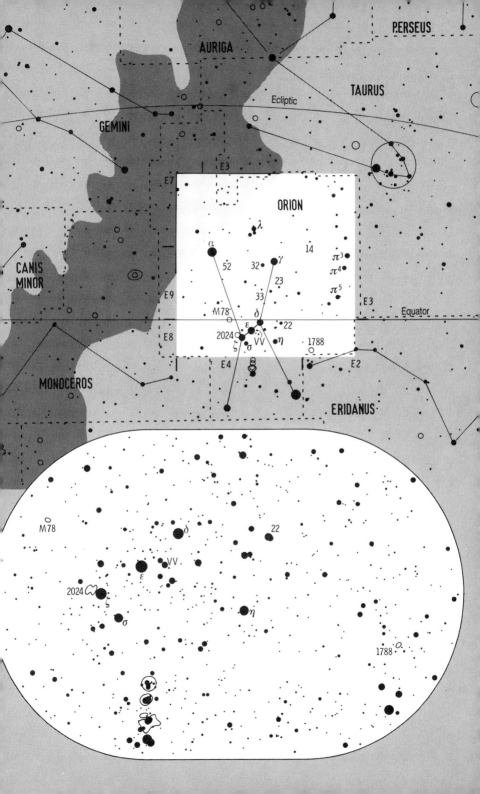

NEBULA	Position	v-Mag.	Size	Shape	Type	Vis.	Dist.	R.A.	Dec.
2287 M41	CMa ▢	5	12/□'	30'	O m	OC	⊞	2500 ly	6ʰ47ᵐ0	−20°73
2362	CMa ▢	4	7	6	O p	OC	⚫	5000	7 18.8	−24.93
2447 M93	Pup ▢	6½	12	15	O m	OC	⊞	4000	7 44.6	−23.87

2287 M41 Nicely resolved in binoculars, an excellent object for small scopes, even visible with the unaided eye as a glow, not too impressive in a telescope, contains few faint stars, a binary is on the northwest side.

2362 ..... In binoculars the mag. 4.4 star τ CMa almost outshines faint background glow of mag. 6 (9 $^m_{□'}$); well resolved in a telescope, triangular.

2447 M93 Binoculars resolve a few bright stars within a nebulous background; a telescope resolves fainter stars well; three parallel chains of stars.

STAR		Position	V-Mag.	B−V	Te.	Abs.	Name	Dist.	R.A.	Dec.
1	ζ CMa	▢ •	3.0 ⚹	−.2	↓	−2ᴹ	. . Phurud	350,800 ly	6ʰ20ᵐ3	−30°06
2	β CMa	▢ ●	2.0	−.2	↓	−4	. . Mirzam . .	500	6 22.7	−17.96
7	ν² CMa	▢ •	4.0	1.0	↓	2	. . . .	65	6 36.7	−19.26
9	α CMa	▢ ●	−1.5	0.0	↓	1	. . Sirius . .	8.6	6 45.1	−16.72
13	κ CMa	▢ •	3.5–4.0⚹	−.1	↓	−4	. . . .	800	6 49.8	−32.51
16	o¹ CMa	▢ •	3.8–4.0	1.7	↓	−5	. . . .	2000	6 54.1	−24.18
21	ε CMa	▢ ●	1.5 ⚹	−.2	↓	−4	. . Adhara . .	430	6 58.6	−28.97
22	σ CMa	▢ •	3.5	1.7	↓	−4	. . . .	1200	7 01.7	−27.93
24	o² CMa	▢ •	3.0	−.1	↓	−7	. . . .	2500	7 03.0	−23.83
23	γ CMa	▢ •	4.1	−.1	↓	−1	. . . .	400	7 03.8	−15.63
25	δ CMa	▢ ●	1.8	0.7	↓	−7	. . Wezen . .	2000	7 08.4	−26.39
27	CMa	▢ •	4.4–4.7	−.2	↓	−4	. EW CMa .	1500	7 14.3	−26.35
28	ω CMa	▢ •	3.8–4.0	−.1	↓	−4	. . . .	1000	7 14.8	−26.77
145	CMa	▢ •	4.5 ⚹	1.1	↓	−4	 2000,250		7 16.6	−23.31
29	CMa	▢ •	4.8–5.3	−.1	↓	−7	. UW CMa .	5000	7 18.7	−24.56
31	η CMa	▢ ●	2.4 ⚹	−.1	↓	−7	. Aludra	2500,600	7 24.1	−29.30
n	Pup	▢ ·	5.1 ⚹	0.4	↓	3	. . . .	95	7 34.3	−23.47
k	κ Pup	▢ •	3.8 ⚹	−.2	↓	−2	. . . .	450	7 38.8	−26.80
3	Pup	▢ •	3.9	0.2	↓	−7	. . . .	5000	7 43.8	−28.96
7	ξ Pup	▢ ●	3.2 ⚹	1.1	↓	−5	Aspidiske	1200,350	7 49.3	−24.86
11	Pup	▢ •	4.2	0.7	↓	−2	. . . .	500	7 56.9	−22.88
15	ϱ Pup	▢ ●	2.8	0.5	↓	1	. . . .	63	8 07.5	−24.30

BINARY		Position	V-Mag.		B−V		Te.	Sep.	PA	Vis.		VARIABLE	STAR
1	ζ CMa	▢ •	3.0	7.7	−.2	1.1	↓↓	176″	•	⊞		13 κ CMa ▢ •	irregular
13	κ CMa	▢ •	4	6.8	−.2	−.1	‖	265.4	••	⊠		16 o¹ CMa ▢ •	irregular
21	ε CMa	▢ ●	1.5	7.5	−.2	0.1	‖	7.5	•	⊙		27 EW CMa ▢ •	irregular
145	CMa	▢ •	4.8	6.0	1.7	0.3	↓↓	27	••	◔		28 ω CMa ▢ •	irregular
31	η CMa	▢ ●	2.4	6.9	−.1	0.0	‖	179	••	⊞		Extrema 3.6–4.2	
n	Pup	▢ ·	5.8	5.9	0.4	0.4	↓↓	9.8	••	◔		29 UW CMa ▢ · ◠◡◠	
k	κ Pup	▢ •	4.5	4.6	−.2	−.1	‖	9.9	••	◔		Period 4.3934 d	
7	ξ Pup	▢ ●	3.3	5.3	1.2	0.8	↓↓	288	••	⊠		Min. 2451202.0	

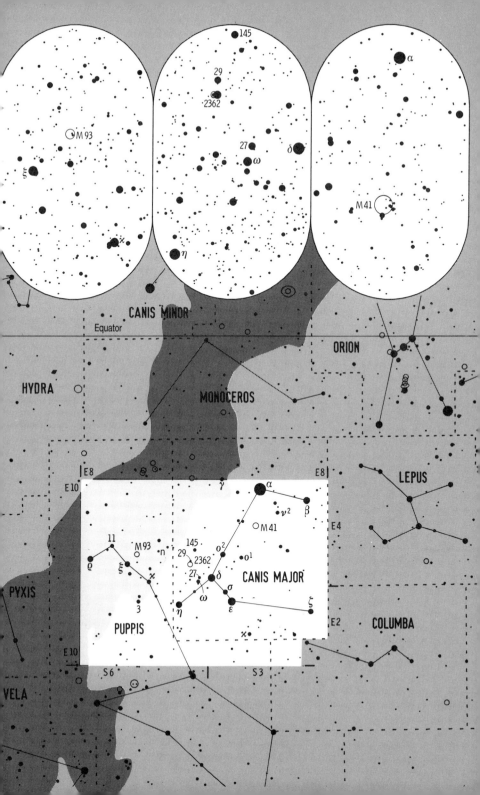

NEBULA	Position	v-Mag.	Size	Shape	Type	Vis.	Dist.	R.A.	Dec.
2129	Gem ⟦⟧	7	10/□′ 5′	○ p	OC	⟦⟧	6000 ly	6ʰ01ᵐ0	23°30
2168 M35	Gem ⟦⟧	5	12	30 ○ r	OC	⟦⟧	3000	6 08.9	24.33
2175	Ori ⟦⟧	7	13	20 ○ p n	OC	⟦⟧	7000	6 09.8	20.32
2261	Mon ⟦⟧	9½	10	1.8 0 Re	DN	⟦⟧	3000	6 39.2	8.73
2264	Mon ⟦⟧	4	9	15 0 p n	OC	⟦⟧	1000	6 41.1	9.88
2392	Gem ⟦⟧	9	8	0.8 ○ D	PN	⟦⟧	2500	7 29.2	20.91

2129 ..... Recognizable as a cluster in a telescope, very sparse, inconspicuous.

2168 M35 Near the limit of the unaided eye, bright glow with some stars in binoculars, nicely resolved in a telescope, impressive at low power.

2175 ..... Very inconspicuous; 10′ north is the dim diffuse nebula NGC 2174.

2261 ..... **Hubble's Variable Nebula**, variable within days, some detail visible in a telescope at high power; it almost looks like a comet.

2264 ..... **Christmas Tree**, elongated, one mag. 4.7 star, others mag. 8–10.

2392 ..... **Eskimo Nebula**, bright green disk, irregularly bright central region; the mag. 10.5 central star is clearly visible at high power.

STAR			Position	V-Mag.	B–V	Te.	Abs.	Name	Dist.	R.A.	Dec.
7	η	Gem ⟦⟧	●	3.2–3.4	1.6	↓	−2ᴹ	. Tejat Prior .	350 ly	6ʰ14ᵐ9	22°51
13	μ	Gem ⟦⟧	●	2.9	1.6	↓	−1	Tejat Posterior	230	6 23.0	22.51
18	ν	Gem	●	4.1	−.1	↓	−2	. . . .	450	6 29.0	20.21
24	γ	Gem	●	1.9	0.0	↓	−1	. . Alhena . .	105	6 37.7	16.40
15		Mon ⟦⟧	●	4.6 ✶	−.2	↓	−3	in NGC 2264	1000	6 41.0	9.90
27	ε	Gem	●	3.1	1.4	↓	−4	. Mebsuta .	900	6 43.9	25.13
31	ξ	Gem ⟦⟧	●	3.4	0.4	↓	2	. . . .	57	6 45.3	12.90
34	ϑ	Gem	●	3.6	0.1	↓	0	. . . .	195	6 52.8	33.96
38		Gem	•	4.7 ✶	0.3	↓	2	. . . .	90	6 54.6	13.18
43	ζ	Gem	●	3.6–4.2✶	0.9	↓	−4	. Mekbuda	1000 ,90	7 04.1	20.57
54	λ	Gem	●	3.6	0.1	↓	1	. . . .	94	7 18.1	16.54
55	δ	Gem ⟦⟧	●	3.5	0.4	↓	2	. . Wasat . .	59	7 20.1	21.98
60	ι	Gem	●	3.8	1.0	↓	1	. . . .	125	7 25.7	27.80
62	ϱ	Gem	•	4.2	0.3	↓	3	. . . .	60	7 29.1	31.78
66	α	Gem	●	1.6 ✶	0.0	↓	1	. . Castor . .	52	7 34.6	31.89
69	υ	Gem	•	4.1	1.5	↓	0	. . . .	240	7 35.9	26.90
77	κ	Gem	●	3.6	0.9	↓	0	. . . .	145	7 44.4	24.40
78	β	Gem	●	1.1	1.0	↓	1	. . Pollux . .	33.5	7 45.3	28.03

BINARY		Position	V-Mag.		B–V		Te.	Sep.	PA	Vis.
15	Mon ⟦⟧	•	4.7	7.7	−.2	0.0	↓↓	2″9	●	⟦⟧
38	Gem	•	4.7	7.7	0.3	0.7	↓↓	7.3	●	⟦⟧
43	ζ Gem	•	4	7.6	0.9	0.6	↓↓	101	●	⟦⟧
66	α Gem	●	1.9	2.9	0.0	0.1	↓↓ ′0	3.9	•●	⟦⟧

2020 ↗ 1995 ● ●

2005	4.3	•● ⟦⟧
2010	4.7	•● ⟦⟧
2015	5.0	•● ⟦⟧

VARIABLE STAR

7 η Gem ⟦⟧ ● semireg.
Period 232.9 d
Min. ≈ 2451230
Extrema 3.2–3.9

43 ζ Gem ⟦⟧ • ⌐‾⌐
Period 10.1508 d
Max. 2451205.8

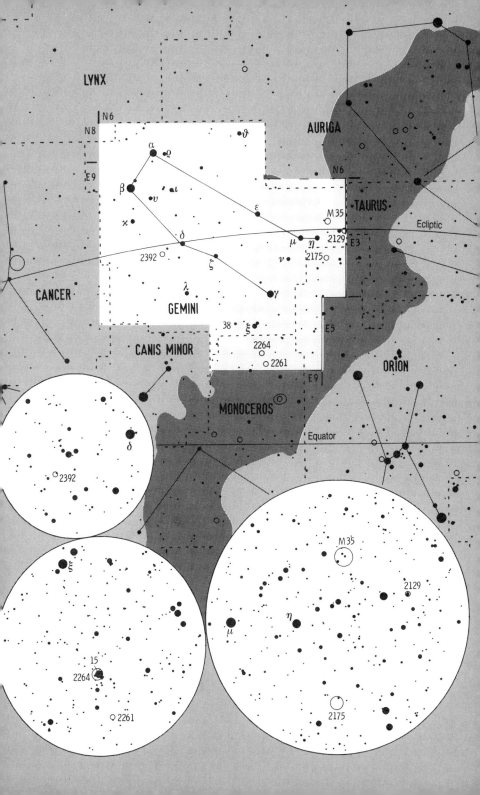

NEBULA	Position	v-Mag.	Size	Shape	Type	Vis.	Dist.	R.A.	Dec.
2323 M 50	Mon	6½ 12/□'	15'	O r	OC		3 500 ly	7ʰ03ᵐ.2	−8°.33
2360	CMa	7½ 12	12	O m	OC		5 000	7 17.8	−15.58
2359	CMa	9 13	10	0 Em	DN		4 000	7 18.0	−13.20
2422 M 47	Pup	4½ 11	25	O m	OC		1 800	7 36.6	−14.50
2423	Pup	7 13	20	O m	OC		3 000	7 37.1	−13.87
2437 M 46	Pup	6 13	25	O r	OC		6 000	7 41.8	−14.82
2438	Pup	11 11	1.0	O R	PN		4 000	7 41.8	−14.74
2539	Pup	7 13	20	O m	OC		4 000	8 10.7	−12.8'

2323 M 50		Brightest stars resolved in binoculars, fully resolved in a telescope, leaves best impression at low power, quite asymmetric, dark center.
2360		A glow in binoculars, even in a telescope not completely resolved, distinct central elongated core, asymmetric shape, chains of stars.
2359		Contains a few stars, diffuse nebula is near the limit of binoculars, oval in a telescope, interesting detail visible through a nebula filter.
2422 M 47		Impressive cluster in binoculars, no better in a telescope, visible with the unaided eye as a dim glow, contains mostly bright stars.
2423		Consists of faint stars, some of which are binaries, not resolved in binoculars, quite symmetric, low contrast to the rich background.
2437 M 46		Bright large oval glow in binoculars, impressive number of stars in a telescope, very rich in faint stars, uniform distribution of stars.
2438		In northern part of M 46, dim, needs high power or a nebula filter.
2539		Difficult in binoculars, excellent in a telescope, several stellar condensations, irregular circumference; it contains about 100 stars.

STAR	Position	V-Mag.	B−V	Te.	Abs.	Name	Dist.	R.A.	Dec.
5 γ Mon		4.0	1.3		-3^M		650 ly	6ʰ14ᵐ.9	−6°.27
11 β Mon		3.7 ☆	−.1		−3		650	6 28.8	−7.03
14 ϑ CMa		4.1	1.4		0		240	6 54.2	−12.04
18 μ CMa		5.0 ☆	1.2		−2		800	6 56.1	−14.04
22 δ Mon		4.2	0.0		−1		370	7 11.9	−0.49
U Mon		5.6−7.6	1.0		−5		3 000	7 30.8	−9.78
26 α Mon		3.9	1.0		1		145	7 41.2	−9.55
2 Pup		5.7 ☆	0.1		0		350	7 45.5	−14.69
5 Pup		5.5 ☆	0.5		3		100	7 47.9	−12.19
29 ζ Mon		4.4	1.0		−5		2 000	8 08.6	−2.98
19 Pup		4.7	0.9		1	near NGC 2539	185	8 11.3	−12.93

BINARY	Position	V-Mag.		B−V		Te.	Sep.	PA	Vis.
11 β Mon		3.8 ☆	7.6	−.1	0.0		248″.0		
		4.5 ☆	4.6	−.1	−.1		8.3		
		5.1	5.4	−.1	−.1		3.0		
18 μ CMa		5.1	7.4	1.4	0.1		2.8		
2 Pup		6.1	6.9	0.1	0.3		16.8		
5 Pup		5.7	7.4	0.5	0.7		1.4		

VARIABLE STAR

U Mon — semireg. Period 92 d. Pulsating star as Mira type stars, light curve similar to β Lyrae type.

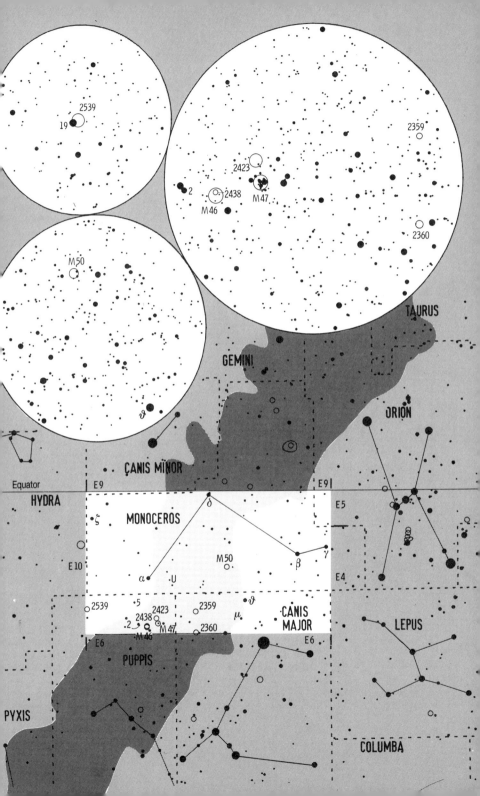

2539
19

2359

2423
2
2438
M47
M46
M50

2360

TAURUS

GEMINI

ORION

ϑ

CANIS MINOR

Equator   E9   E9

HYDRA

MONOCEROS   δ   E5

ζ

β   γ

M50

E10   α   ·U   E4

2539   ·5   ϑ
         2438 2423   2359   μ   CANIS
      2   M47   2360   MAJOR

E6   E6
M46
PUPPIS

LEPUS

PYXIS

COLUMBA

NEBULA	Position		v-Mag.	Size		Shape	Type	Vis.	Dist.	R.A.	Dec.
2237	Mon	⬡	6	14/□'	80'	O Em	DN	⬡	5 000 ly	6ʰ32.ᵐ3	5.03
2244	Mon	⬡	5	12	25	O p n	OC	⬡	5 000	6 32.4	4.87
2301	Mon	⬡	6	11	15	0 m	OC	⬡	2 500	6 51.8	0.47
2324	Mon	⬡	8½	13	7	O r	OC	⬡	12 000	7 04.2	1.07
2632 M 44	Cnc	⬡	3½	13	80	O m	OC	⬡	580	8 40.1	19.98
2682 M 67	Cnc	⬡	7	13	20	O r	OC	⬡	2 500	8 50.8	11.82

2237 ..... **Rosette Nebula**, nebula filter recommended (NGC 2237–39, 2246).
2244 ..... In Rosette Nebula, fine in binoculars, no better in a telescope.
2301 ..... Partially resolved in binoculars, completely resolved in a telescope, conspicuous chains of stars arranged in the direction north–south.
2324 ..... Well visible only in a telescope, shows an impressive number of stars.
2632 M 44 **Praesepe, Beehive**, easily visible with the unaided eye as a glow, impressive in binoculars, for a telescope with a wide field of view.
2682 M 67 Large nebula in binoculars, beautifully resolved in a telescope.

STAR	Position			V-Mag.	B–V	Te.	Abs.	Name	Dist.	R.A.	Dec.
8 ε	Mon	⬡	•	4.3	✶ 0.2	↓	1ᴹ	....	128 ly	6ʰ23.ᵐ8	4.59
T	Mon	⬡	·	5.6–6.6	1.0	↓	−6	....	6 000	6 25.2	7.09
3 β	CMi	▫	•	2.9	−.1	↓	−1	. Gomeisa .	170	7 27.2	8.29
4 γ	CMi	▫	•	4.3	1.4	↓	−1	....	400	7 28.2	8.93
10 α	CMi	▫	●	0.4	0.4	↓	3	.. Procyon ..	11.4	7 39.3	5.22
16 ζ	Cnc	▫	•	4.7	✶ 0.5	↓	3	....	84	8 12.2	17.65
17 β	Cnc	▫	•	3.5	1.5	↓	−1	.. Altarf ..	300	8 16.5	9.19
23 φ²	Cnc	▫	·	5.5	✶ 0.2	↓	1	....	280	8 26.8	26.94
43 γ	Cnc	⬡	•	4.7	0.0	↓	1	Asellus Borealis	160	8 43.3	21.47
47 δ	Cnc	⬡	•	3.9	1.1	↓	1	Asellus Australis	135	8 44.7	18.15
48 ι	Cnc	▫	•	3.9	✶ 0.9	↓	−1	....	300	8 46.7	28.76
55 ϱ¹	Cnc	▫	•	5.3	✶ 1.1	↓	−1	55, 53 Cnc	41,800	8 52.5	28.30
57	Cnc	▫	·	5.4	✶ 1.0	↓	0	....	370	8 54.2	30.58
X	Cnc	⬡	·	6.0–6.5	3.2	•	−3	....	2 000	8 55.4	17.23
65 α	Cnc	▫	•	4.3	0.1	↓	1	. Acubens .	175	8 58.5	11.86
RS	Cnc	▫	·	5.3–6.3	1.5	↓	0	....	420	9 10.6	30.96

BINARY	Position			V-Mag.		B–V		Te.	Sep.	PA	Vis.		VARIABLE STAR	
8 ε	Mon	⬡	•	4.4	6.6	0.2	0.4	↓↓	12.″3		⬡		T Mon ⬡	╱‾
16 ζ	Cnc	▫	•	5.0	✶ 6.1	0.5	0.7	↓↓	6.2		⬡		Period	27.025 d
				5.6	6.0	0.5	0.6	↓↓'0	0.9		⬡		Max.	2451216
									2005	1.0	⬡		Min.	Max.+20
									2010	1.1	⬡		X Cnc ⬡	· semireg.
									2015	1.1	⬡		Period	≈ 180 d
23 φ²	Cnc	▫	·	6.2	6.3	0.2	0.2	↓↓	5.2		⬡		Extrema	5.6–7.5
48 ι	Cnc	▫	·	4.0	6.5	1.0	0.1	↓↓	30.5		⬡		Very orange star.	
55 ϱ¹	Cnc	▫	·	6.0	6.3	0.9	1.5	↓↓	273		⬡		RS Cnc ▫	· semireg.
57	Cnc	▫	·	6.0	6.3	1.0	1.1	↓↓	1.5		⬡		Period	≈ 120 d

(inset diagram for 16 ζ Cnc: "→ 2020", "1995·", with dots)

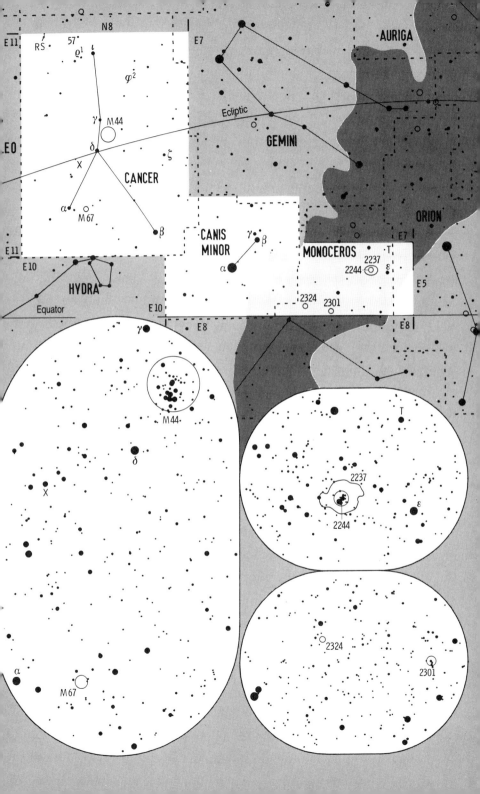

NEBULA	Position		v-Mag.	Size		Shape	Type	Vis.	Dist.	R.A.	Dec.
2548 M 48	Hya	◐	6	13/□′	40′	0 m	OC	⊞	2 200 ly	8ʰ13ᵐ8	−5°80
3115	Sex	◒	9½	11	5	⫽ S0	Glx	⊡	25 M	10 05.2	−7.72
3242	Hya	◓	8	6	0.6	O D	PN	⊡	3 000	10 24.8	−18.64

2548 M 48   Fine bright open cluster in binoculars, not much better in a telescope, bright stars and binaries form a central bar aligned north-south; Messier listed its declination as 5° further to the north.

3115 .....   **Spindle Galaxy**, spindle shape only recognizable in a telescope at high power; it is a fine edge-on galaxy with a bright elongated core.

3242 .....   **Ghost of Jupiter**, similar size and shape as Jupiter, stellar in binoculars, distinct blue-green disk in a telescope, needs high power, high surface brightness; a bright knot lies at its northwest edge.

STAR	Position			V-Mag.	B−V	Te.	Abs.	Name	Dist.	R.A.	Dec.
C	Hya	◐	•	3.9	0.0	↓	1ᴹ	....	125 ly	8ʰ25ᵐ7	−3°91
4 δ	Hya	•	•	4.1	0.0	↓	0	....	175	8 37.7	5.70
5 σ	Hya	•	•	4.4	1.2	↓	−1	....	370	8 38.8	3.34
β	Pyx	•	•	4.0	0.9	↓	−1	....	380	8 40.1	−35.31
7 η	Hya	•	•	4.3	−.2	↓	−1	....	370	8 43.2	3.40
α	Pyx	•	•	3.7	⚹ −.2	↓	−3	....	850	8 43.6	−33.19
11 ε	Hya	•	•	3.4	⚹ 0.7	↓	0	} Sep. 43′ •	135	8 46.8	6.42
13 ϱ	Hya	•	•	4.4	0.0	↓	−1		340	8 48.4	5.84
γ	Pyx	•	•	4.0	1.3	↓	0	....	210	8 50.5	−27.71
16 ζ	Hya	•	•	3.1	1.0	↓	0	....	150	8 55.4	5.95
22 ϑ	Hya	•	•	3.9	−.1	↓	1	....	125	9 14.4	2.31
27	Hya	•	•	4.7	⚹ 0.8	↓	0	....	225	9 20.5	−9.56
30 α	Hya	•	●	2.0	1.4	↓	−2	.. Alphard ..	180	9 27.6	−8.66
31 τ¹	Hya	•	•	4.5	⚹ 0.5	↓	3	....	57	9 29.1	−2.77
35 ι	Hya	•	•	3.9	1.3	↓	−1	....	270	9 39.9	−1.14
39 υ¹	Hya	•	•	4.1	0.9	↓	−1	....	280	9 51.5	−14.85
8 γ	Sex	•	·	5.1	0.0	↓	1	....	260	9 52.5	−8.11
15 α	Sex	•	·	4.5	0.0	↓	0	....	280	10 07.9	−0.37
41 λ	Hya	◒	•	3.6	1.0	↓	1	....	115	10 10.6	−12.35
42 μ	Hya	◒	•	3.8	1.5	↓	−1	....	250	10 26.1	−16.84
α	Ant	•	•	4.3	1.4	↓	−1	....	360	10 27.2	−31.07
U	Hya	•	· 4.7–5.1		3.0	·	−1	....	500	10 37.6	−13.38
35	Sex	•	·	5.8	⚹ 1.2	↓	−1	....	700	10 43.3	4.75
ν	Hya	•	●	3.1	1.2	↓	0	....	138	10 49.6	−16.19

BINARY	Position		V-Mag.		B−V	Te.	Sep.	PA	Vis.
α Pyx	•	•	3.7	8.0	−.2	0.9 ↓↓	106″6	⦂	⊙
11 ε Hya	•	•	3.4	6.9	0.7	0.6 ↓↓	2.9	•⦁	⊙
27 Hya	•	·	4.8	7.0	0.9	0.4 ↓↓	229.2	•⦁	⊠
31 τ¹ Hya	•	·	4.6	7.2	0.4	0.9 ↓↓	65.7	⦂	⊙
35 Sex	•	·	6.1	7.2	1.2	1.0 ↓↓	6.8	•⦁	⊙

VARIABLE STAR

U   Hya · semireg.
Period   115 d
Max. ≈ 2451275
Third of reddest stars in catalog.

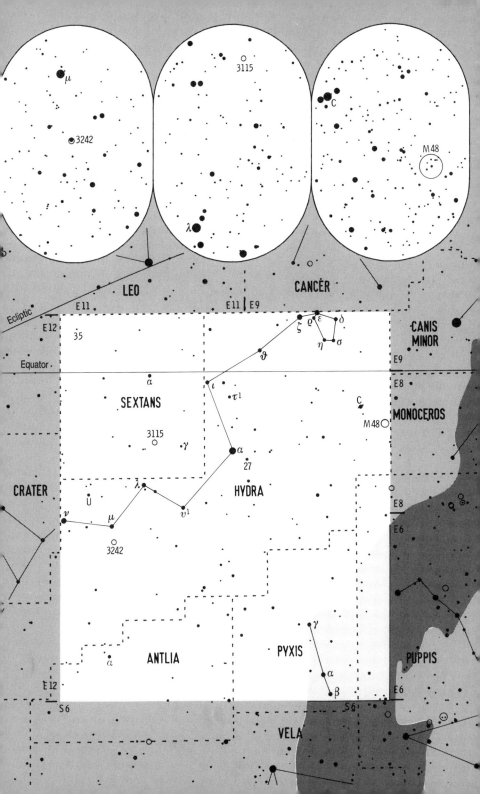

NEBULA	Position		v-Mag.	Size		Shape	Type	Vis.	Dist.	R.A.	Dec.
2903	Leo	○	9	13/□′	10′	0	Sc	Glx	⟨○○⟩	25 M ly	9ʰ32ᵐ2	21°50
3351 M95	Leo	○	10	12	4	0	Sb	Glx	⟨·•⟩	40 M	10 44.0	11.70
3368 M96	Leo	○	9½	12	5	0	Sa	Glx	⟨·•⟩	40 M	10 46.8	11.82
3379 M105	Leo	○	9½	12	3	O	E1	Glx	⟨·•⟩	40 M	10 47.8	12.58
3384	Leo	○	10	12	4	0	S0	Glx	⟨·•⟩	40 M	10 48.3	12.63
3623 M65	Leo	○	9½	12	8	I	Sa	Glx	⟨·•⟩	40 M	11 18.9	13.09
3627 M66	Leo	○	9	12	6	0	Sb	Glx	⟨·•⟩	40 M	11 20.2	13.00
3628	Leo	○	10	12	12	I	Sb	Glx	⟨·•⟩	40 M	11 20.3	13.59

2903	Galaxy with bright oval center, asymmetric, relatively easy to find.
3351 M95	Stellar core, arms of barred spriral not detectable, 41′ west of M96.
3368 M96	Elongated halo and central area; it contains a bright stellar core.
3379 M105	Stellar core, more easily visible than M95; it is 48′ north of M96.
3384	Lies only 8′ east of M105, stellar core within a featureless nebula.
3623 M65	Circular central region within a very elongated asymmetric halo.
3627 M66	At the limit of visibility of binoculars; it is an interesting object in a telescope due to dark irregular dust features; the core is elongated.
3628	Nicely elongated, a faint dust lane lies along the southern edge.

STAR	Position		V-Mag.		B−V	Te.	Abs.	Name	Dist.	R.A.	Dec.
2 ω Leo	⟨·•⟩	·	5.4	☆	0.6	↓	3ᴹ	....	112 ly	9ʰ28ᵐ5	9°06
4 λ Leo	⟨○⟩	•	4.3		1.5	↓	−1	.. Alterf ..	320	9 31.7	22.97
14 o Leo	⟨·•⟩	•	3.5		0.5	↓	0	....	134	9 41.2	9.89
17 ε Leo	⟨○⟩	•	3.0		0.8	↓	−2	....	260	9 45.9	23.77
R Leo	⟨·•⟩	·	5.8–10		1.4	↓	1	....	300	9 47.6	11.43
24 μ Leo	⟨·•⟩	•	3.9		1.2	↓	1	....	134	9 52.8	26.01
30 η Leo	⟨·•⟩	•	3.5		0.0	↓	−6	....	2000	10 07.3	16.76
32 α Leo	⟨·•⟩	●	1.4	☆	−.1	↓	−1	.. Regulus ..	77	10 08.4	11.97
36 ζ Leo	⟨·•⟩	•	3.4		0.3	↓	−1	. Aldhafera .	260	10 16.7	23.42
41 γ Leo	⟨·•⟩	●	2.0	☆	1.1	↓	−1	.. Algieba ..	125	10 20.0	19.84
47 ϱ Leo	⟨·○⟩	•	3.8		−.1	↓	−6	....	3000	10 32.8	9.31
54 Leo	⟨·•⟩	•	4.3	☆	0.0	↓	0	....	290	10 55.6	24.75
68 δ Leo	⟨·•⟩	•	2.6		0.1	↓	1	.. Zosma ..	58	11 14.1	20.52
70 ϑ Leo	⟨○⟩	•	3.3		0.0	↓	0	... Coxa ...	170	11 14.2	15.43
78 ι Leo	⟨·○⟩	•	4.0	☆	0.4	↓	2	....	80	11 23.9	10.53
94 β Leo	⟨·•⟩	●	2.1		0.1	↓	2	. Denebola .	36	11 49.1	14.57

BINARY	Position		V-Mag.		B−V		Te.	Sep.	PA	Vis.
2 ω Leo	⟨·•⟩	·	5.9	6.5	0.6	0.6	↓↓′0	0″6	••	○
								2015	0.8	••	⟨•⟩
32 α Leo	⟨·•⟩	●	1.4	7.9	−.1	0.9	↓↓	175.9	••	⟨··⟩
41 γ Leo	⟨·•⟩	●	2.3	3.5	1.1	1.1	↓↓	4.7	••	⟨·•⟩
54 Leo	⟨·•⟩	•	4.5	6.3	0.0	0.1	↓↓	6.6	••	⟨·•⟩
78 ι Leo	⟨·○⟩	•	4.1	6.7	0.4	0.6	↓↓′0	1.7	••	⟨•⟩
								2015	2.1	••	⟨•⟩

VARIABLE STAR

R Leo ⟨·⟩ · ⟍⟋

Period	≈ 312 d
Max.	≈ 2451360
Min.	Max. + 180
Extrema	4.4–11.3

The period varies by a few days.

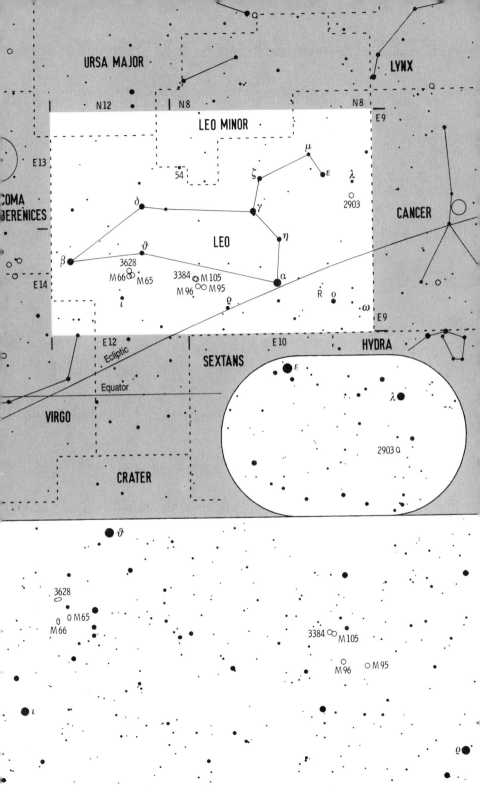

NEBULA	Position		v-Mag.	Size	Shape	Type	Vis.	Dist.	R.A.	Dec.
4361	Crv	⊙	10½ 11/□′	1.2	O D	PN	⊙	4000 ly	12ʰ24ᵐ.5	−18°.79
4590 M68	Hya	⊙	8 13	10	O X	GC	⊙	30000	12 39.5	−26.74
4594 M104	Vir	⊙	8½ 12	8	] Sa	Glx	⊙	50 M	12 40.0	−11.62
4697	Vir	⊙	9½ 11	2.5	] E6	Glx	⊙	60 M	12 48.6	−5.80

4361 ..... Faint planetary, requires high power; the central star is only mag. 13.
4590 M68 Resolved only in a telescope, but then even in the very center.
4594 M104 **Sombrero Galaxy**, very elongated, spindle shape barely visible in binoculars, impressive in a telescope, dust lane nearly right through the center, small double core; a chain of stars lies 25′ to the west.
4697 ..... Small, elongated, contains a stellar nucleus, otherwise featureless.

STAR		Position		V-Mag.	B−V	Te.	Abs.	Name	Dist.	R.A.	Dec.
7	α	Crt	⊙ •	4.1	1.1	↓	0ᴹ	.. Alkes ..	180 ly	10ʰ59ᵐ.8	−18°.30
11	β	Crt	⊙ •	4.5	0.0	↓	0	....	260	11 11.7	−22.83
12	δ	Crt	⊙ •	3.6	1.1	↓	0	....	200	11 19.3	−14.78
15	γ	Crt	⊙ •	4.1	0.2	↓	2	....	84	11 24.9	−17.68
84	τ	Leo	⊙ •	4.9 ⚹	0.9	↓	−2	 600,1000		11 27.9	2.85
N		Hya	⊙ •	4.9 ⚹	0.5	↓	3	....	87	11 32.3	−29.26
ξ		Hya	⊙ •	3.5	0.9	↓	1	....	130	11 33.0	−31.86
3	ν	Vir	⊙ •	4.0	1.5	↓	−1	....	300	11 45.9	6.53
5	β	Vir	⊙ •	3.6	0.5	↓	3	. Zawijava .	35.5	11 50.7	1.76
β		Hya	⊙ •	4.3	−.1	↓	−1	....	360	11 52.9	−33.91
1	α	Crv	⊙ •	4.0	0.3	↓	3	.. Alchiba ..	49	12 08.4	−24.73
2	ε	Crv	⊙ •	3.0	1.3	↓	−2	....	300	12 10.1	−22.62
4	γ	Crv	⊙ •	2.6	−.1	↓	−1	.. Gienah ..	165	12 15.8	−17.54
15	η	Vir	⊙ •	3.9	0.0	↓	−1	.. Zaniah ..	260	12 19.9	−0.67
7	δ	Crv	⊙ •	2.9	0.0	↓	1	.. Algorab ..	88	12 29.9	−16.52
9	β	Crv	⊙ •	2.7	0.9	↓	−1	....	140	12 34.4	−23.40
26	χ	Vir	⊙ •	4.6	1.2	↓	0	....	310	12 39.2	−8.00
29	γ	Vir	⊙ •	2.7 ⚹	0.4	↓	2	.. Porrima..	39	12 41.7	−1.45
46	γ	Hya	⊙ •	3.0	0.9	↓	0	....	132	13 18.9	−23.17
R		Hya	⊙	· 4.9–9.0	1.6	↓	−2	....	800	13 29.7	−23.28

BINARY		Position		V-Mag.		B−V	Te.	Sep.	PA	Vis.
84	τ	Leo	⊙ •	5.0	7.5	1.0	0.4	↓↓	89″	⊙⊙
N		Hya	⊙ •	5.6	5.8	0.5	0.5	↓↓	9.5	⊙
29	γ	Vir	⊙ •	3.5	3.5	0.4	0.4	↓↓′0	1.5	⊙

	Sep.	Vis.
2002	1.2	⊙
2004	0.8	⊙
2006	0.4	○
2008	0.8	⊙
2010	1.3	⊙
2012	1.7	⊙
2015	2.2	⊙

(orbit diagram: 2020 ... 1995)

VARIABLE STAR

R Hya ⊙ ·

Period 387 d
Max. 2451330
Min. Max.+200
Extrema 3.5–10.9
The period has been decreasing; it was close to 500 days during the early 1700s.

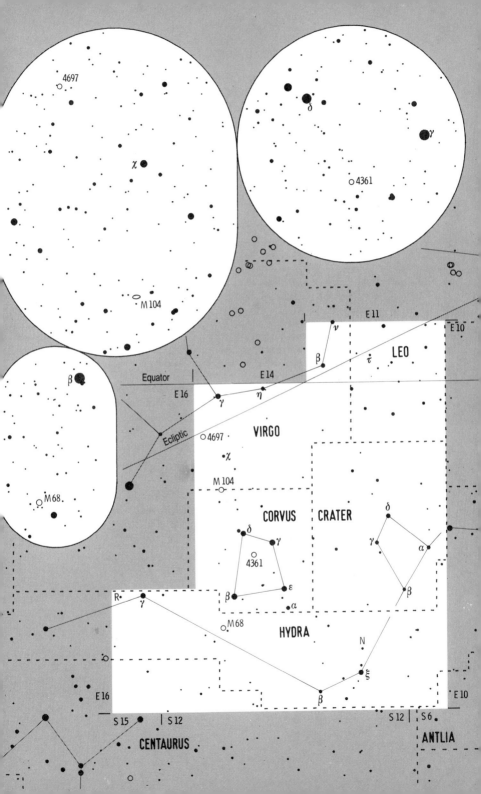

NEBULA	Position		v-Mag.	Size	Shape	Type	Vis.	Dist.	R.A.	Dec.
Coma Cluster	Com	:o	2½	13/□′	300′	O p	**OC**	⊙	290 ly	12^{h}25^m	26°0
4494	Com	:o	10	11	2	O E2	**Glx**	•	60 M	12 31.4	25.77
4559	Com	:o	10	13	8	0 Sd	**Glx**	o°	40 M	12 36.0	27.96
4565	Com	:o	10	13	15	∣ Sb	**Glx**	o°	60 M	12 36.3	25.99
4631	CVn	⊙	9½	12	12	∣ Sd	**Glx**	••	35 M	12 42.1	32.54
4656	CVn	⊙	10½	13	12	∣ Sm	**Glx**	o	35 M	12 44.0	32.17
4725	Com	:o	9½	14	9	0 Sa	**Glx**	o°	60 M	12 50.4	25.50
4826 M 64	Com	:•	9	12	6	0 Sa	**Glx**	⊙	22 M	12 56.7	21.68
5024 M 53	Com	⁰	8	12	7	O V	**GC**	∷	60 000	13 12.9	18.17

Coma Cluster Distinctly visible with unaided eye under dark sky counting about 10 stars, fine in opera glasses, almost too big for binoculars, completely inconspicuous in a telescope, see comment at bottom right.

4494 ..... Quite bright core within a uniform nebulosity, almost circular.

4559 ..... Distinctly elongated; central region has a slightly asymmetric shape.

4565 ..... Wonderful edge-on galaxy, impressively long in a telescope, huge true size, nearly central dust lane in the bright part of the galaxy, irregular brightness distribution in central area, thin extensions.

4631 ..... Edge-on galaxy like NGC 4565, but no dust lane, many irregular knots and asymmetries visible in a telescope, worthwhile object.

4656 ..... Difficult edge-on galaxy, very long, modestly bright central region; the other condensation 3′ northeast of the core is called NGC 4657.

4725 ..... Dark sky and low power necessary, small faint core visible in a telescope, but not the bar; the halo has extremly low surface brightness.

4826 M 64 **Black Eye Galaxy**, elongated dark dust feature next to the core just visible in a telescope, distinct elongated central region south of the geometric center; the outer outline is sharp unlike most galaxies.

5024 M 53 Distinct core; the outer regions are partially resolved in a telescope.

STAR	Position		V-Mag.	B–V	Te.	Abs.	Name	Dist.	R.A.	Dec.
2	Com	•	5.8 ✳	0.2	↓	1^M		350 ly	12^{h}04^{m}3	21°46
15 γ	Com	:o	4.4	1.1	↓	1		170	12 26.9	28.27
17	Com	:o	5.0 ✳	0.0	↓	0 in Coma Cluster		280	12 28.9	25.91
24	Com	•	4.8 ✳	0.9	↓	−1		500	12 35.1	18.38
32,33	Com	•	5.8 ✳	1.1	↓	−2	1 500,400		12 52.3	17.09
35	Com	:o	4.9 ✳	0.9	↓	0		330	12 53.3	21.24
42 α	Com	⁰	4.3	0.5	↓	3 .. Diadem ..		50	13 10.0	17.53
43 β	Com	o	4.2	0.6	↓	4		30	13 11.9	27.88

BINARY	Position		V-Mag.		B–V		Te.	Sep.	PA	Vis.
2	Com	•	6.1	7.5	0.2	0.3	↓↓	3″7	••	•
17	Com	:o	5.3	6.6	0.0	0.2	↓↓	145.2	••	⊞
24	Com	•	5.0	6.5	1.1	0.3	↓↓	20.1	••	•
32,33	Com	•	6.3	6.9	1.6	0.6	↓↓	196	••	⊞
35	Com	:o	5.0	7.2	1.0	0.4	↓↓	1.1	•	•

COMA CLUSTER

1) The group of stars visible by unaided eye.

2) This name may also refer to a distant cluster of galaxies.

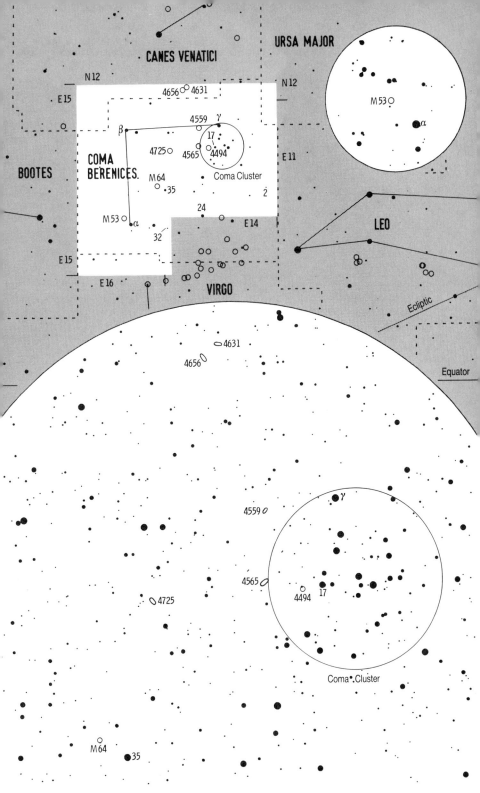

NEBULA	Position		v-Mag.	Size	Shape		Type	Vis.	Dist.	R.A.	Dec.
4192 M 98	Com		10½	13/□'	7'	[Sa	Glx	⊙	60 M ly	$12^h13^m.8$	14°.90
4216	Vir		10½	13	8	‖ Sb	Glx	⊙	60 M	12 15.9	13.15
4254 M 99	Com		10	13	4	O Sc	Glx	⊙	60 M	12 18.8	14.42
4303 M 61	Vir		10	12	4	O Sc	Glx	•	60 M	12 21.9	4.47
4321 M 100	Com		9½	13	5	O Sc	Glx	⊙	60 M	12 22.9	15.82
4374 M 84	Vir		9½	12	3	O E1	Glx	•	60 M	12 25.1	12.89
4382 M 85	Com		9½	12	4	0 S0	Glx	•	60 M	12 25.4	18.19
4406 M 86	Vir		9½	12	4	0 E3	Glx	•	60 M	12 26.2	12.95
3 C 273	Vir		13	(8)	<0.1	O Quasar		⊙	2500 M	12 29.1	2.05
4472 M 49	Vir		8½	12	5	O E2	Glx	⊙	60 M	12 29.8	8.00
4486 M 87	Vir		9	12	4	O E1	Glx	⊙	60 M	12 30.8	12.39
4501 M 88	Com		10	13	6	0 Sb	Glx	⊙	60 M	12 32.0	14.42
4526	Vir		10	11	4	‖ S0	Glx	•	60 M	12 34.0	7.70
4548 M 91	Com		10½	13	3.5	O Sb	Glx	⊙	60 M	12 35.4	14.50
4552 M 89	Vir		10	11	2	O E0	Glx	•	60 M	12 35.7	12.56
4569 M 90	Vir		10	13	8	‖ Sa	Glx	⊙	60 M	12 36.8	13.16
4579 M 58	Vir		10	13	5	O Sb	Glx	⊙	60 M	12 37.7	11.82
4621 M 59	Vir		10	11	2.5	0 E5	Glx	•	60 M	12 42.0	11.65
4649 M 60	Vir		9	11	3	O E2	Glx	•	60 M	12 43.7	11.55
4762	Vir		10½	11	5	‖ S0	Glx	•	60 M	12 52.9	11.23

The **Virgo Cluster** is the nearest of the rich clusters of galaxies. The central region is marked as the dashed oval in the chart at bottom. The whole area of the cluster includes galaxies in the charts N10–N14 and E10–E14. Near the center, faint galaxies are so abundant that it is hard to find one's way around.

4192 M 98 Distinctly elongated with faint diffuse halo, core not outstanding.

4216 ..... Fine, remarkable edge-on galaxy although very faint, stellar core.

4254 M 99 Bright central area, light patches and a hint of the southern spiral arm; 40' east are the magnitude 12 galaxies NGC 4298 and 4302.

4303 M 61 Bright core; the spiral arms are only barely visible in a telescope.

4321 M 100 Elongated central area with a stellar nucleus; the halo is uniform.

4374 M 84 Featureless glow; 16' south is elongated mag. 11½ galaxy NGC 4388.

4382 M 85 Bright central area; 8' east is mag. 11 companion galaxy NGC 4394.

4406 M 86 Featureless; **Makarian's Galaxy Chain** to M 88, galaxies mag. 11.

3 C 273 Brightest **Quasar**, probably an active, especially bright nucleus of a galaxy at enormous distance, visible as a very faint stellar dot.

4472 M 49 The brightest galaxy of the Virgo Cluster, a large luminous galaxy.

4486 M 87 **Virgo A**, central galaxy of the Virgo Cluster, bright round core.

4501 M 88 Asymmetric halo, faint detail; **Makarian's Galaxy Chain** to M 86.

4526 ..... Asymmetric, nucleus not centered in the galaxy, takes high power.

4548 M 91 Featureless; Messier's M 91 not uniquely identified as NGC 4548.

4552 M 89 Small circular galaxy, a round glow with a bright, nearly stellar core.

4569 M 90 Bright central elongated area, largest galaxy of the Virgo Cluster.

4579 M 58 The bar of the barred spiral is just barely visible in a telescope.

4621 M 59 Elongated halo, round central area, stellar core, medium power best.

4649 M 60 Intense stellar core; 3' northwest is mag. 11½ galaxy NGC 4647.

4762 ..... Faint spindle, elongated core; 11' northwest is mag. 11 NGC 4754.

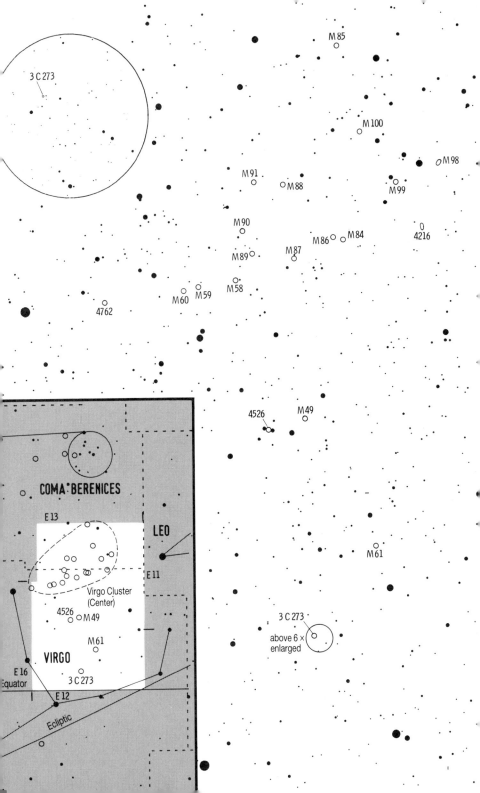

3 C 273

M 85

M 100

M 98

M 91

M 88

M 99

M 90

4216

M 86 ○ M 84

M 89 ○

M 87

M 58

M 60 ○ M 59

4762

4526

M 49

COMA BERENICES

E 13

LEO

Virgo Cluster
(Center)

E 11

4526 ○ M 49

M 61

M 61

M 87

VIRGO

3 C 273

above 6 ×
enlarged

E 16

Equator

3 C 273

E 12

Ecliptic

E15 —— Equator, Ecliptic —— Spring-Summer Constellations

NEBULA	Position	v-Mag.	Size	Shape	Type	Vis.	Dist.	R.A.	Dec.
5272 M 3	CVn	6½ 11/□′	10′	O VI	GC	▣	30 000 ly	13ʰ42ᵐ.2	28°.38
5746	Vir	11 12	6	| Sb	Glx	⊙	90 M	14 44.9	1.95
5904 M 5	Ser	6 11	12	O V	GC	▣	25 000	15 18.6	2.08

5272 M 3   Bright globular cluster, but hard to find, especially outer parts resolved in a telescope, rectangular central area with the small core west of the geometric center, oval halo, curved radial chains of stars.

5746 .....   Very faint, almost edge-on, furthest galaxy of this catalog; the mag. 3.7 star 109 Virginis simplifies finding but interferes with observing.

5904 M 5   Excellent, especially in a telescope, relatively easily resolved, several dense areas and chains of stars, asymmetric elliptical halo; this is the brightest globular cluster of the northern celestial hemisphere.

STAR		Position	V-Mag.	B−V	Te.	Abs.	Name	Dist.	R.A.	Dec.
5	υ Boo	▣ ●	4.1	1.5	↓	0ᴹ	. . . .	250 ly	13ʰ49ᵐ.5	15°.80
8	η Boo	▣ ●	2.7	0.6	↓	2	. . Muphrid . .	36.5	13 54.7	18.40
16	α Boo	▣ ●	0.0	1.2	↓	0	. . **Arcturus** . .	36.5	14 15.7	19.18
25	ϱ Boo	▣ ●	3.6	1.3	↓	0	. . . .	155	14 31.8	30.37
29	π Boo	▣ ●	4.5 ✶	0.0	↓	0	. . . .	320	14 40.7	16.42
30	ζ Boo	▣ ●	3.8	0.0	↓	0	. . . .	180	14 41.1	13.73
34	Boo	▣	4.7–4.9	1.7	↓	−2	. . W Bootis . .	800	14 43.4	26.53
36	ε Boo	▣ ●	2.4 ✶	1.0	↓	−2	Izar, Pulcherrima	210	14 45.0	27.07
109	Vir	▣ ●	3.7	0.0	↓	1	. . . .	130	14 46.2	1.89
37	ξ Boo	▣ ●	4.5 ✶	0.7	↓	5	. . . .	21.9	14 51.4	19.10
2	η CrB	▣ ●	5.0 ✶	0.6	↓	4	. . . .	59	15 23.2	30.29
3	β CrB	▣ ●	3.7	0.3	↓	1	. . Nusakan . .	116	15 27.8	29.11
4	ϑ CrB	▣ ●	4.1	−.1	↓	−1	. . . .	310	15 32.9	31.36
5	α CrB	▣ ●	2.2	0.0	↓	0	**Alphekka, Gemma**	75	15 34.7	26.71
8	γ CrB	▣ ●	3.8 ✶	0.0	↓	1	. . . .	145	15 42.7	26.30
R	CrB	▣ ●	5.7–6.3	0.7	↓	−5	. . . .	4000	15 48.6	28.16
10	δ CrB	▣ ●	4.6	0.8	↓	1	. . . .	165	15 49.6	26.07
13	ε CrB	▣ ●	4.1	1.2	↓	0	. . . .	230	15 57.6	26.88

BINARY		Position	V-Mag.		B−V	Te.	Sep.	PA	Vis.
29	π Boo	▣ ●	4.9	5.8	−.1	0.2	⇈	5″.5	•◦ ⊡
36	ε Boo	▣ ●	2.5	4.9	1.1	0.1	↓↓	2.9	• ⊙
37	ξ Boo	▣ ●	4.7	6.9	0.7	1.2	↓↓′0	6.7	•′ ⊡
				1995			2007	6.3	•′ ⊡
				2020			2015	5.7	•′ ⊡
2	η CrB	▣ ●	5.6	5.9	0.6	0.6	↓↓′0	0.7	′• ⊙
			1995				2007	0.5	•◦ ◯
				2020			2015	0.6	•‥ ◯
8	γ CrB	▣ ●	4.1	5.5	0.0	0.1	↓↓′0	0.7	•◦ ◯
							2007	0.7	‥• ◯
		1995	2020				2015	0.5	•◦ ◯

VARIABLE STAR

34 W Boo ▣ • semireg.
Period 30–450 d
Extrema 4.7–5.4

R CrB ▣ · irregular
Extrema 5.7–14.8

R CrB stars are variables staying usually near maximum light with a rapid decrease and slow increase.

80

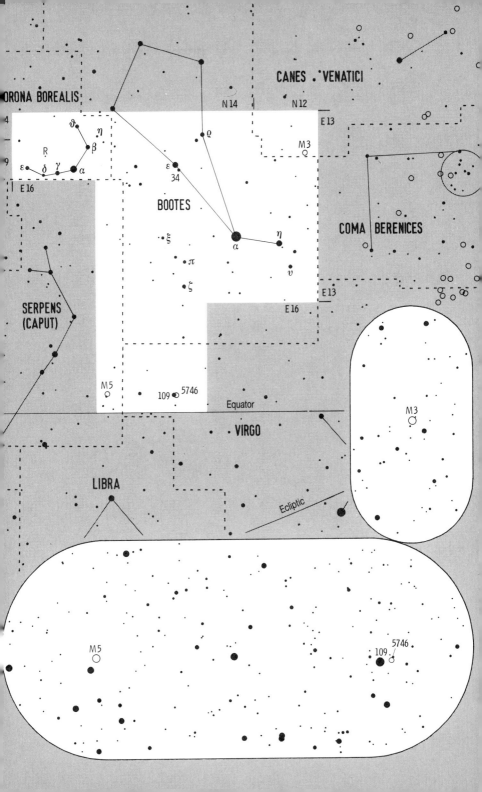

NEBULA	Position	v-Mag.	Size	Shape	Type	Vis.	Dist.	R.A.	Dec.
5236 M83	Hya 💡	8	12/□′	8′	○ Sc	**Glx** 📳	20 Mly	13ʰ37ᵐ.0	−29°.87

5236 M83 Easy object for binoculars, beautiful in a telescope, bright condensed core in an elongated bar with traces of two spiral arms, large halo.

STAR		Position	V-Mag.	B−V	Te.	Abs.	Name	Dist.	R.A.	Dec.
43	δ	Vir	• 3.4	1.6	↓ −1ᴹ	. . . .		200 ly	12ʰ55ᵐ.6	3°.40
47	ε	Vir	• 2.8	0.9	↓ 0	Vindemiatrix	103	13 02.2	10.96
51	ϑ	Vir	• 4.4	0.0	↓ −1	. . . .	400	13 09.9	−5.54
67	α	Vir	● 1.0	−.2	↓ −4	. . **Spica** . .	260	13 25.2	−11.16
79	ζ	Vir	• 3.4	0.1	↓ 2	. . . .	74	13 34.7	−0.60
3		Cen	• 4.3 ⚹	−.1	↓ −1	. . . .	320	13 51.8	−32.99
93	τ	Vir	• 4.2	0.1	↓ 0	. . . .	220	14 01.6	1.54
49	π	Hya	● 3.3	1.1	↓ 1	. . . .	104	14 06.4	−26.68
107	μ	Vir	• 3.9	0.4	↓ 3	. . . .	61	14 43.1	−5.66
54		Hya	• 5.0 ⚹	0.3	↓ 3	. . . .	100	14 46.0	−25.44
7	μ	Lib	• 5.3 ⚹	0.1	↓ 1	. . . .	230	14 49.3	−14.15
9,8	α	Lib	● 2.6 ⚹	0.2	↓ 1	Zubenelgenubi	78	14 50.9	−16.04
19	δ	Lib	• 4.9–5.9	0.0	↓ 0	. . . .	310	15 01.0	−8.52
20	σ	Lib	• 3.3	1.7	↓ −1	. . . .	280	15 04.1	−25.28
27	β	Lib	● 2.6	−.1	↓ −1	Zubeneschemali	160	15 17.0	−9.38
13	δ	Ser	• 3.8 ⚹	0.3	↓ 0	. . . .	210	15 34.8	10.54
38	γ	Lib	• 3.9	1.0	↓ 1	. . . .	150	15 35.5	−14.79
39	υ	Lib	• 3.6	1.4	↓ 0	. . . .	200	15 37.0	−28.13
40	τ	Lib	• 3.7	−.2	↓ −2	. . . .	450	15 38.7	−29.78
24	α	Ser	● 2.6	1.2	↓ 1	. Unukalhai .	73	15 44.3	6.43
28	β	Ser	• 3.7	0.1	↓ 0	. . . .	155	15 46.2	15.42
35	κ	Ser	• 4.1	1.6	↓ −1	. . . .	350	15 48.7	18.14
32	μ	Ser	• 3.5	0.0	↓ 0	. . . .	155	15 49.6	−3.43
R		Ser	· 6.0–13	1.4	↓ −2	. . . .	1000	15 50.7	15.13
37	ε	Ser	• 3.7	0.1	↓ 2	. . . .	70	15 50.8	4.48
5	χ	Lup	• 4.0	0.0	↓ 0	. . . .	210	15 51.0	−33.63
41	γ	Ser	• 3.9	0.5	↓ 4	. . . .	36.5	15 56.5	15.66
	ξ	Lup	• 4.6 ⚹	0.1	↓ 1	. . . .	210	15 56.9	−33.97
	ξ	Sco	• 4.1 ⚹	0.5	↓ 2	. . . .	100	16 04.4	−11.37

BINARY		Position	V-Mag.		B−V		Te.	Sep.	PA	Vis.
3	Cen	•	4.6	6.1	−.1	0.0	↓↓	7″.8	••	⊡
54	Hya	•	5.1	7.1	0.3	0.6	↓↓	8.2	••	⊡
7	μ Lib	•	5.7	6.7	0.0	0.2	↓↓	2.0	⦂	⊙
9,8	α Lib	•	2.8	5.2	0.2	0.4	↓↓	231.1	••	⊠
13	δ Ser	•	4.2	5.2	0.3	0.3	↓↓	4.0	⦂	⊡
	ξ Lup	•	5.1	5.6	0.1	0.1	↓↓	10.2	••	⊡
	ξ Sco	•	4.1⚹	6.9	0.5	0.8	↓↓	280.4	⦂	⊠
			4.2	7.3	0.5	0.7	↓↓	7.7	••	⊡

VARIABLE STAR

19 δ Lib • · 〰

Period 2.32736 d
Min. 2451201.90
Eclipse 12 hours

R Ser · 〰

Period 356 d
Max. 2451208
Min. Max.+210

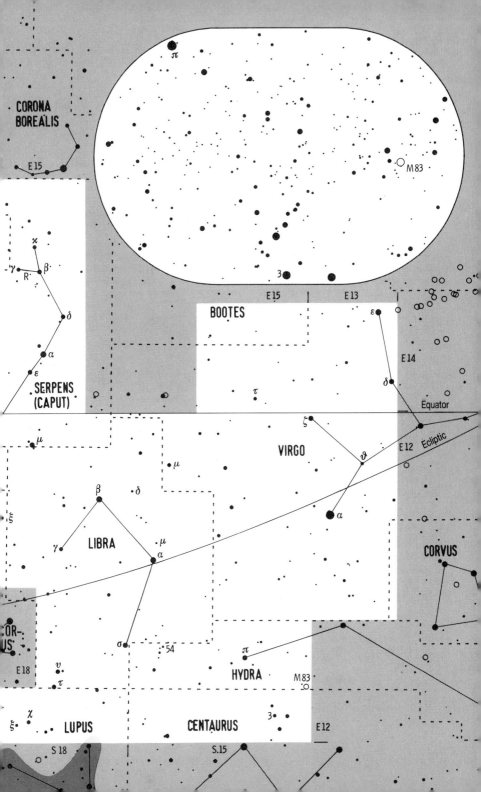

CORONA
BOREALIS

E 15

π

M 83

SERPENS
(CAPUT)

χ

γ
R
β

δ

α
ε

E 15   E 13

3

BOOTES

ε

E 14

δ

Equator

ζ

Ecliptic

τ

VIRGO

ϑ   E 12

μ

μ

α

δ

β

α   CORVUS

LIBRA   μ
γ   α

OR-
US

ξ

σ   54

π

E 18

υ

τ

HYDRA   M 83

χ

3

ξ   LUPUS   CENTAURUS   E 12

S 18   S.15

NEBULA	Position	v-Mag.	Size		Shape	Type	Vis.	Dist.	R.A.	Dec.
6171 M 107	Oph 🔵	8½	12/□'	6'	O X	GC	🔵	20 000 ly	16ʰ32.ᵐ5	−13.°06
6218 M 12	Oph 🔵	7	12	12	O IX	GC	🔵	18 000	16 47.2	−1.95
6254 M 10	Oph 🔵	7	12	12	O VII	GC	🔵	15 000	16 57.1	−4.10
6333 M 9	Oph 🔵	8	11	5	O VIII	GC	🔵	25 000	17 19.2	−18.52
6402 M 14	Oph 🔵	8	12	8	O VIII	GC	🔵	30 000	17 37.6	−3.24
IC 4665 ...	Oph 🔵	4½	13	50	O p	OC	🔵	1 000	17 46.3	5.72

6171 M 107	Very difficult to resolve even in a telescope, uniform, oval halo.
6218 M 12	Slightly elliptical glow in binoculars, well resolved in a telescope, looks similar to some rich open clusters, chains of stars in the halo.
6254 M 10	Outer region well resolved in a telescope, nebulous background, oval.
6333 M 9	Barely resolvable, similar globular cluster NGC 6356 is 1° northeast.
6402 M 14	Oval featureless nebula in a telescope, not resolvable into stars.
IC 4665 ...	Conspicuous in opera glasses or binoculars, but not in a telescope.

STAR	Position	V-Mag.	B−V	Te.	Abs.	Name	Dist.	R.A.	Dec.
1 δ	Oph 🔵 ●	2.7	1.6	↓	−1ᴹ	. Yed Prior .	170 ly	16ʰ14.ᵐ3	−3.°69
2 ε	Oph 🔵 ●	3.2	1.0	↓	1	Yed Posterior	109	16 18.3	−4.69
7 χ	Oph 🔵 ●	4.2–4.7	0.2	↓	−2	....	500	16 27.0	−18.46
10 λ	Oph 🔵 ●	3.8 ✳	0.0	↓	0	.. Marfik ..	170	16 30.9	1.98
13 ζ	Oph 🔵 ●	2.5	0.0	↓	−3	....	460	16 37.2	−10.57
27 κ	Oph 🔵 ●	3.2	1.2	↓	1	....	86	16 57.7	9.38
35 η	Oph 🔵 ●	2.4	0.1	↓	0	.. Sabik ..	83	17 10.4	−15.72
55 α	Oph 🔵 ●	2.1	0.1	↓	1	. Rasalhague .	47	17 34.9	12.56
55 ξ	Ser 🔵 ●	3.5	0.3	↓	1	....	105	17 37.6	−15.40
60 β	Oph 🔵 ●	2.8	1.2	↓	1	.. Cebalrai ..	82	17 43.5	4.57
61	Oph 🔵 ·	5.6 ✳	0.1	↓	0	....	500	17 44.6	2.58
62 γ	Oph 🔵 ●	3.8	0.0	↓	1	....	95	17 47.9	2.71
Barnard	🔵 ✝	9.5	1.6	↓	13	Barnard's Star	5.94	17 57.8	4.69
64 ν	Oph 🔵 ●	3.3	1.0	↓	0	....	150	17 59.0	−9.77
67	Oph 🔵 ●	3.9 ✳	0.0	↓	−5	....	2 000	18 00.6	2.93
69 τ	Oph 🔵 ·	4.8 ✳	0.4	↓	1	....	170	18 03.1	−8.18
70	Oph 🔵 ●	4.0 ✳	0.9	↓	5	....	16.6	18 05.5	2.50
72	Oph 🔵 ●	3.7 ✳	0.2	↓	1	......	83,400	18 07.4	9.56

BINARY	Position	V-Mag.		B−V		Te.	Sep.	PA	Vis.
10 λ	Oph 🔵 ●	4.2	5.2	0.0	0.1	↓↓	1."5		🔵
61	Oph 🔵 ·	6.2	6.6	0.1	0.1	↓↓	20.7		🔵
67	Oph 🔵 ●	4.0	8.1	0.0	0.0	↓↓	54.4		🔵
69 τ	Oph 🔵 ·	5.2	5.9	0.4	0.4	↓↓	1.7		🔵
70	Oph 🔵 ●	4.2	6.0	0.8	1.1	↓↓'0	3.7		🔵
						2005	4.9		🔵
				1995		2010	5.7		🔵
		2020				2015	6.3		🔵
72	Oph 🔵 ●	3.7	7.5	0.1	1.1	↓↓	287		🔵

VARIABLE STAR

7 χ Oph 🔵 • irregular

Note: Barnard's Star
It is the star with the largest proper motion of 10."4/year, also the nearest star in the northern hemisphere, distance 2015: 5.93 ly.

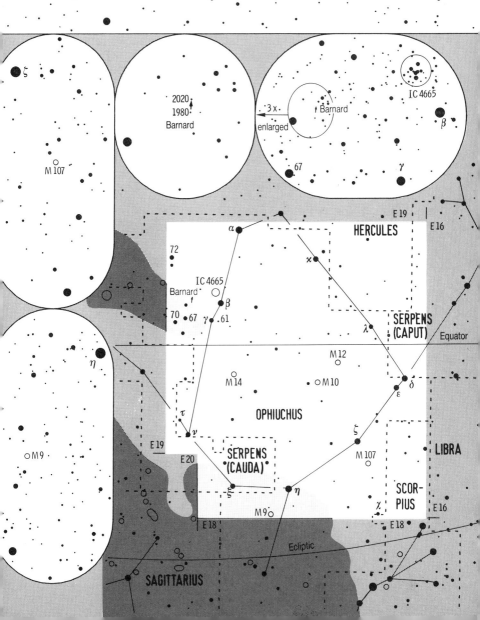

NEBULA		Position		v-Mag.	Size	Shape	Type	Vis.	Dist.	R.A.	Dec.
6093	M 80	Sco		7½ 11/□′	5′	○ II	GC	⬡	30 000 ly	16^{h}17^{m}0	−22°98
6121	M 4	Sco		6 12	18	○ IX	GC	⬡	7 000	16 23.6	−26.53
6266	M 62	Oph		7 11	8	○ IV	GC	⬡	20 000	17 01.2	−30.11
6273	M 19	Oph		7 11	8	○ VIII	GC	⬡	30 000	17 02.6	−26.27
6369	.....	Oph		11 9	0.5	○ R	PN	⦿	4 000	17 29.3	−23.76
6405	M 6	Sco		4½ 10	20	◖ m	OC	⬡	1 800	17 40.1	−32.22
6475	M 7	Sco		3½ 12	50	○ m	OC	⬡	900	17 53.9	−34.82

6093	M 80	Very bright central area, takes high power, but hardly resolvable.
6121	M 4	Easy to find, beautifully resolved in a telescope, central stellar bar.
6266	M 62	Distinctly asymmetric, nebulous arms, interesting globular cluster.
6273	M 19	Quite oval, edges are resolvable into stars, asymmetric, large core.
6369	.....	Difficult, stellar at low power, a disk at high power, hardly a ring.
6405	M 6	**Butterfly Cluster**, an excellent object for every scope, elongated.
6475	M 7	Easily visible by unaided eye, nicely resolved in opera glasses, not better in a telescope, irregular; it is southernmost Messier object.

STAR			Position	V-Mag.	B−V	Te.	Abs.	Name	Dist.	R.A.	Dec.
5	ϱ	Sco	•	3.9	−.2		−2^M	....	450 ly	15^{h}56^{m}9	−29°21
6	π	Sco	•	2.9	−.2		−3	....	450	15 58.9	−26.11
7	δ	Sco	•	2.3	−.1		−3	....	450	16 00.3	−22.62
8	β	Sco	•	2.4 ✻	−.1		−4	.. Acrab ..	600	16 05.4	−19.80
9	ω¹	Sco	•	3.9	0.0		−2	....	420	16 06.8	−20.67
14	ν	Sco	•	3.9 ✻	0.0		−2	....	420	16 12.0	−19.46
20	σ	Sco	•	2.9	0.2		−4	....	800	16 21.2	−25.59
5	ϱ	Oph	•	4.4 ✻	0.2		−1	....	420	16 25.6	−23.44
21	α	Sco	●	0.9–1.1✻	1.8		−5	.. Antares ..	450	16 29.4	−26.43
23	τ	Sco	•	2.8	−.2		−3	....	450	16 35.9	−28.22
26	ε	Sco	•	2.3	1.1		1	....	65	16 50.2	−34.29
RR		Sco	•	6.0–10	1.3		−2	....	1 000	16 56.6	−30.58
36		Oph	•	4.3 ✻	0.9		5	....	19.5	17 15.3	−26.60
39	o	Oph	•	4.9 ✻	0.9		0	....	350	17 18.0	−24.29
42	ϑ	Oph	•	3.3	−.2		−3	....	550	17 22.0	−25.00

BINARY			Position	V-Mag.		B−V		Te.	Sep.	PA	Vis.
8	β	Sco	•	2.6	4.9	−.1	0.0	‖	13.7		⦿
14	ν	Sco	•	4.0✻	6.3✻	0.0	0.1	‖	41.0		⦿
	A			4.4	5.4	0.0	0.0	‖	1.4		⦿
	B			6.7	7.8	0.1	0.2	‖	2.6		⦿
5	ϱ	Oph		4.6✻	6.8	0.2	0.3	‖	156.3		⬡
			″	7.3		″	0.3		151.1		⬡
				5.0	5.7	0.2	0.2	‖	3.0		⦿
21	α	Sco	●	1	5.5	1.9	0.0		2.8		⦿
36		Oph	•	5.1	5.1	0.9	0.9		4.9		⦿
39	o	Oph	•	5.1	6.6	1.0	0.5		10.1		⦿

VARIABLE STAR

21 α Sco	● semireg.
Period	4–5 years
Extrema	0.9–1.8
Color	contrast
with	companion!

RR Sco

Period	277 d
Max.	2451265
Min.	Max. + 150
Extrema	5.0–12.4

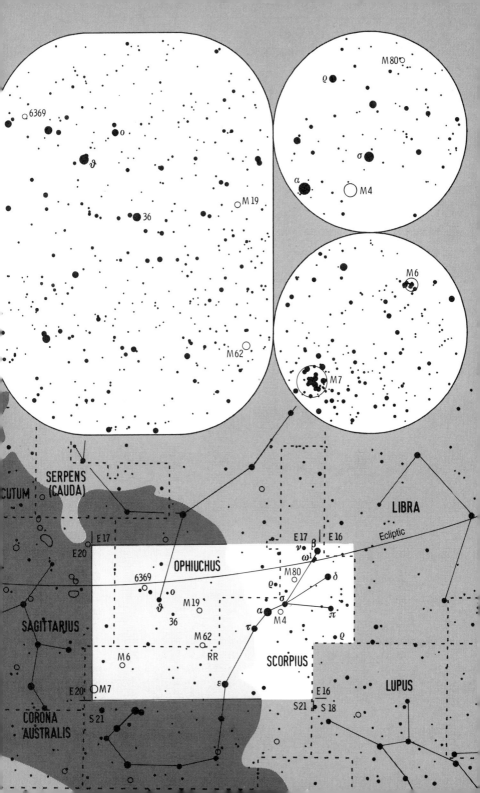

E19 ——— Equator, Ecliptic ——— Summer Constellations

NEBULA	Position		v-Mag.	Size	Shape	Type	Vis.	Dist.	R.A.	Dec.
6210	Her	[icon]	9	6/□′	0.3	O D	PN	[icon]	5000 ly	16ʰ44.ᵐ5	23°.80
6572	Oph	[icon]	8½	5	0.2	O D	PN	[icon]	2000	18 12.1	6.86
6633	Oph	[icon]	5	10	20	0 m	OC	[icon]	1000	18 27.7	6.57
IC 4756 ...	Ser	[icon]	5	13	60	O m	OC	[icon]	1400	18 39.0	5.43
6694 M 26	Sct	[icon]	8	12	8	O p	OC	[icon]	5000	18 45.2	−9.40
6705 M 11	Sct	[icon]	6	11	12	O r	OC	[icon]	6000	18 51.1	−6.27
6712	Sct	[icon]	8½	12	5	O IX	GC	[icon]	22000	18 53.1	−8.70

6210 ..... Stellar in binoculars, small blue-green disk in a high power eyepiece.
6572 ..... Stellar except at highest power and in good seeing, color blue-green.
6633 ..... Quite bright, impressive irregular features, a rewarding object.
IC 4756 ... Sparse, only a few scattered stars, best in a finder or in binoculars.
6694 M 26 A faint open cluster, only resolved in a telescope, asymmetric.
6705 M 11 Bright glow in binoculars, slightly triangular, impressive number of stars in a telescope; a distinct mag. 8.4 star is close to the center.
6712 ..... Faint globular cluster, at most a few stars resolved in a telescope.

STAR		Position	V-Mag.		B−V	Te.	Abs.	Name	Dist.	R.A.	Dec.
7 κ	Her	[icon] •	4.7	✳	1.0	↓	−1ᴹ	....	400 ly	16ʰ08.ᵐ1	17°.05
20 γ	Her	[icon] •	3.7		0.3	↓	0	....	200	16 21.9	19.15
27 β	Her	[icon] ●	2.8		0.9	↓	−1	. Ruticulus .	160	16 30.2	21.49
64 α	Her	[icon] ●	2.6–3.4	✳	1.3	↓	−3	. Rasalgethi .	400	17 14.6	14.39
65 δ	Her	[icon] ●	3.1		0.1	↓	1	....	78	17 15.0	24.84
86 μ	Her	[icon] ●	3.4		0.8	↓	4	....	27.3	17 46.5	27.72
92 ξ	Her	[icon] ●	3.7		0.9	↓	1	....	135	17 57.8	29.25
95	Her	[icon] •	4.3	✳	0.4	↓	−2	....	500	18 01.5	21.60
103 o	Her	[icon] •	3.8		0.0	↓	−1	....	350	18 07.5	28.76
100	Her	[icon] ·	5.1	✳	0.1	↓	1	....	240	18 07.8	26.10
58 η	Ser	[icon] ●	3.2		0.9	↓	2	....	62	18 21.3	−2.90
109	Her	[icon] ●	3.8		1.2	↓	1	....	130	18 23.7	21.77
59 d	Ser	[icon] •	5.2	✳	0.5	↓	−1	....	500	18 27.2	0.20
α	Sct	[icon] •	3.9		1.3	↓	0	....	175	18 35.2	−8.24
5	Aql	[icon] •	5.6	✳	0.2	↓	1	....	270	18 46.5	−0.96
β	Sct	[icon] ●	4.2		1.1	↓	−3	....	750	18 47.2	−4.75
R	Sct	[icon] ·	5.0–6.5		1.4	↓	−4	....	2000	18 47.5	−5.70
63 ϑ	Ser	[icon] ●	4.0	✳	0.2	↓	1	... Alya ...	140	18 56.2	4.20

BINARY		Position	V-Mag.		B−V	Te.	Sep.	PA	Vis.
7 κ	Her	[icon] •	5.0	6.2	0.9	1.1	↕↕	27″.2	[icon]
64 α	Her	[icon] ●	3–4	5.4	1.5	0.7	↕↕	4.9	[icon]
95	Her	[icon] •	4.9	5.1	0.1	0.9	↕↕	6.4	[icon]
100	Her	[icon] ·	5.8	5.9	0.1	0.2	↕↕	14.2	[icon]
59 d	Ser	[icon] •	5.3	7.6	0.5	0.3	↕↕	3.7	[icon]
5	Aql	[icon] •	5.9	7.5	0.1	0.3	↕↕	12.7	[icon]
63 ϑ	Ser	[icon] •	4.6	5.0	0.2	0.2	↕↕	22.4	[icon]

VARIABLE STAR

64 α Her [icon] • semireg.
Perd. 50 d – 6 years
Binary star mag.
2.7–3.6 and 5.4.
R Sct [icon] · semireg.
Period 140–146 d
Extrema 4.2–8.6

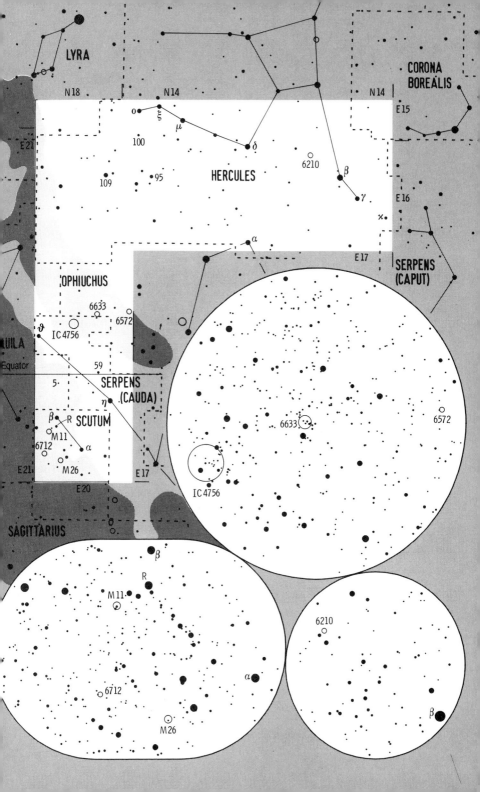

NEBULA	Position		v-Mag.	Size	Shape	Type	Vis.	Dist.	R.A.	Dec.
6494 M 23	Sgr	⬡	6	13/□′	25′	O m	OC	⬡	2 200 ly	17ʰ56ᵐ.8 −19°.02
6514 M 20	Sgr	⬡	7	13	20	O Em	DN	⬡	6 000	18 02.6 −23.03
6523 M 8	Sgr	⬡	4½	13	60	0 Em	DN	⬡	6 000	18 03.8 −24.38
6531 M 21	Sgr	⬡	6½	11	10	O m	OC	⬡	4 000	18 04.6 −22.50
M 24	Sgr	⬡	4	13	100	0 Milky Way		⬡	8 000	18 16.9 −18.48
6611 M 16	Ser	⬡	6	12	25	O Em	DN	⬡	7 000	18 18.8 −13.78
6613 M 18	Sgr	⬡	7	12	10	O p n	OC	⬡	4 000	18 19.9 −17.13
6618 M 17	Sgr	⬡	6	13	35	O Em	DN	⬡	6 000	18 20.8 −16.18
6626 M 28	Sgr	⬡	7	11	6	O IV	GC	⬡	20 000	18 24.5 −24.87
6637 M 69	Sgr	⬡	8	11	4	O V	GC	⬡	30 000	18 31.4 −32.34
IC 4725 M 25	Sgr	⬡	5	12	30	O m	OC	⬡	2 500	18 31.6 −19.23
6656 M 22	Sgr	⬡	5½	11	20	O VII	GC	⬡	10 000	18 36.4 −23.90
6681 M 70	Sgr	⬡	8	11	4	O V	GC	⬡	30 000	18 43.2 −32.29
6715 M 54	Sgr	⬡	8	11	4	O III	GC	⬡	80 000	18 55.1 −30.48

6494 M 23 Resolved in binoculars, impressive in a telescope at low power.
6514 M 20 **Trifid Nebula**, division into three parts by three radial dust bands, structure visible in a telescope at low power through a nebula filter.
6523 M 8 **Lagoon Nebula**, visible to the unaided eye, for every scope, fantastic through a nebula filter, open cluster NGC 6530 in eastern part.
6531 M 21 Resolved in binoculars, few bright stars, inconspicuous, near M 20.
M 24 Messier describes clearly the Milky Way cloud and not NGC 6603.
6611 M 16 **Eagle Nebula**, nebula with dust areas, some 20 stars embedded.
6613 M 18 Sparse, inconspicuous since the surrounding field is quite rich.
6618 M 17 **Omega Nebula, Swan Nebula**, fantastic structure, bright arms, knots, and dark dust clouds, more detail through a nebula filter.
6626 M 28 Asymmetric shape, bright central area; it is barely resolvable.
6637 M 69 Faint, outer region partially resolved in a telescope, irregular outline.
IC 4725 M 25 Very nicely resolved in binoculars, some irregular stellar groups.
6656 M 22 Very bright oval, impressive in a telescope, uncountable stars.
6681 M 70 Rather faint, distinct center, outer portions only just resolvable.
6715 M 54 Not resolvable, bright concentrated core, takes high power well.

STAR	Position		V-Mag.	B−V	Te.	Abs.	Name	Dist.	R.A.	Dec.
10 γ Sgr	•	•	3.0	1.0	↓	1ᴹ	.. Alnasl ..	97 ly	18ʰ05ᵐ.8 −30°.42
13 μ Sgr	⬡	•	3.8	0.2	↓	−7	. . . .	4 000	18 13.8 −21.06
19 δ Sgr	⬡	•	2.7	1.4	↓	−2	Kaus Media	300	18 21.0 −29.83
20 ε Sgr	•	●	1.8	0.0	↓	−1	Kaus Australis	145	18 24.2 −34.38
22 λ Sgr	⬡	•	2.8	1.0	↓	1	Kaus Borealis	78	18 28.0 −25.42
27 φ Sgr	•	•	3.2	−.1	↓	−1	. . . .	230	18 45.7 −26.99
34 σ Sgr	•	●	2.0	−.1	↓	−2	.. Nunki ..	220	18 55.3 −26.30
37 ξ² Sgr	•	•	3.5	1.2	↓	−2	. . . .	350	18 57.7 −21.11
38 ζ Sgr	⬡	●	2.6	0.1	↓	0	. . . .	90	19 02.6 −29.88
39 o Sgr	•	•	3.8	1.0	↓	1	. . . .	140	19 04.7 −21.74
40 τ Sgr	•	•	3.3	1.2	↓	0	. . . .	120	19 06.9 −27.67
41 π Sgr	•	•	2.9	0.4	↓	−3	. . . .	430	19 09.8 −21.02

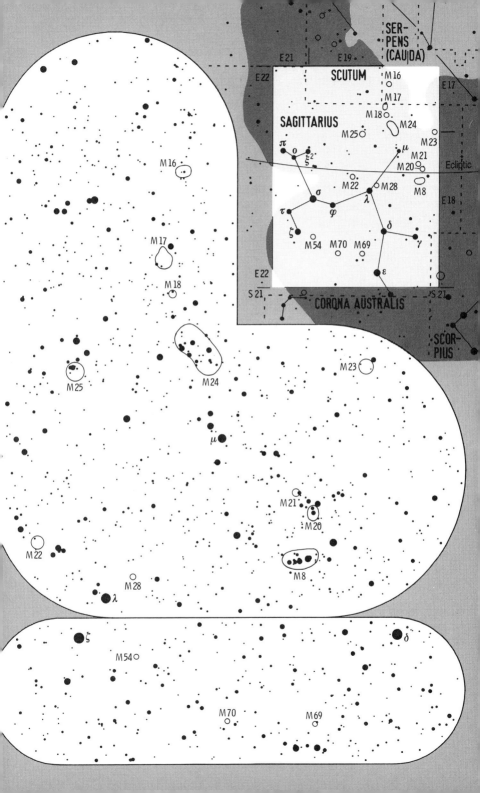

NEBULA	Position	v-Mag.	Size	Shape	Type	Vis.	Dist.	R.A.	Dec.
6838 M 71	Sge [○]	8½	12/□′	5′	O IX	**GC**	☉	13 000 ly	19ʰ53ᵐ8	18°78
6853 M 27	Vul [○]	7	10	7	0 A	**PN**	⊞	1 000	19 59.6	22.72

6838 M 71 Interesting features, triangular shape, resolved into stars in a telescope, low number of stars, looks similar to some open clusters.

6853 M 27 **Dumbbell Nebula**, may be the most beautiful planetary, shape visible in binoculars, more detail in a telescope, greenish color, southwestern lobe is brighter, extended faint halo requires nebula filter.

STAR		Position	V-Mag.	B−V	Te.	Abs.	Name	Dist.	R.A.	Dec.
13	ε Aql	[•] •	4.0	1.1	↓	1ᴹ		150 ly	18ʰ59ᵐ6	15°07
12	Aql	[•] •	4.0	1.1	↓	1		150	19 01.7	−5.74
15	Aql	[•] •	5.2 ✳	1.2	↓	0		330, 600	19 05.0	−4.03
17	ζ Aql	[•] ●	3.0	0.0	↓	1		84	19 05.4	13.86
16	λ Aql	[•] ●	3.4	−.1	↓	0		125	19 06.2	−4.88
R	Aql	[•] ·	5.8–10	1.3	↓	−1		700	19 06.4	8.23
30	δ Aql	[•] ●	3.4	0.3	↓	2		50	19 25.5	3.11
6	α Vul	[•] ●	4.4	1.5	↓	0		300	19 28.7	24.66
5	α Sge	[•] ●	4.4	0.8	↓	−1 } Sep. 35′ ✦		460	19 40.1	18.01
6	β Sge	[•] ●	4.4	1.0	↓	−1 }		460	19 41.0	17.48
50	γ Aql	[•] ●	2.7	1.5	↓	−3 . . Tarazed . .		500	19 46.3	10.61
7	δ Sge	[○] ●	3.7	1.3	↓	−2		460	19 47.4	18.53
52	π Aql	[•] ·	5.7 ✳	0.5	↓	0		500	19 48.7	11.82
53	α Aql	[•] ●	0.8	0.2	↓	2 **Altair, Atair**		16.7	19 50.8	8.87
55	η Aql	[•] ●	3.5–4.4	0.7	↓	−5		1 400	19 52.5	1.01
57	Aql	[•] ·	5.3 ✳	−.1	↓	0		350	19 54.6	−8.23
60	β Aql	[•] ●	3.7	0.9	↓	3 . Alschain .		45	19 55.3	6.41
12	γ Sge	[○] ●	3.5	1.6	↓	−1		260	19 58.8	19.49
16	Vul	[○] ·	5.2 ✳	0.4	↓	1		220	20 02.0	24.94
15	Sge	[•] ·	5.4 ✳	0.5	↓	0		58, 600	20 04.1	17.08
65	ϑ Aql	[•] ●	3.2	−.1	↓	−1		280	20 11.3	−0.82
2	ε Del	[•] ●	4.0	−.1	↓	−1		350	20 33.2	11.30
6	β Del	[•] ●	3.6	0.4	↓	1		100	20 37.5	14.60
9	α Del	[•] ●	3.8	−.1	↓	−1		240	20 39.6	15.91
11	δ Del	[•] ●	4.4	0.3	↓	0		210	20 43.5	15.07
12	γ Del	[•] ●	3.9 ✳	0.8	↓	1		105	20 46.7	16.12

BINARY		Position	V-Mag.		B−V		Te.	Sep.	PA	Vis.
15	Aql	[•] ·	5.4	7.0	1.1	1.5	↓↓	39″	●	[•]
52	π Aql	[•] ·	6.3	6.8	0.8	0.1	↓↓	1.4	·●	[•]
57	Aql	[•] ·	5.7	6.5	−.1	0.0	↓↓	35.7	●	[•]
16	Vul	[○] ·	5.8	6.2	0.3	0.4	↓↓	0.9	·●	[•]
15	Sge	[•] ·	5.8	6.9	0.6	0.1	↓↓	215	●	[⊞]
12	γ Del	[•] ·	4.3	5.1	1.0	0.5	↓↓'0	9.2	●●	[•]
								2015	8.9	●●	[•]

VARIABLE STAR

R Aql	[• ·] ⟨curve⟩
Period	≈ 280 d
Max.	≈ 2451285

55 η Aql	[• ●] ⟨curve⟩
Period	7.1767 d
Max.	2451206.0
Min.	Max. + 4.9

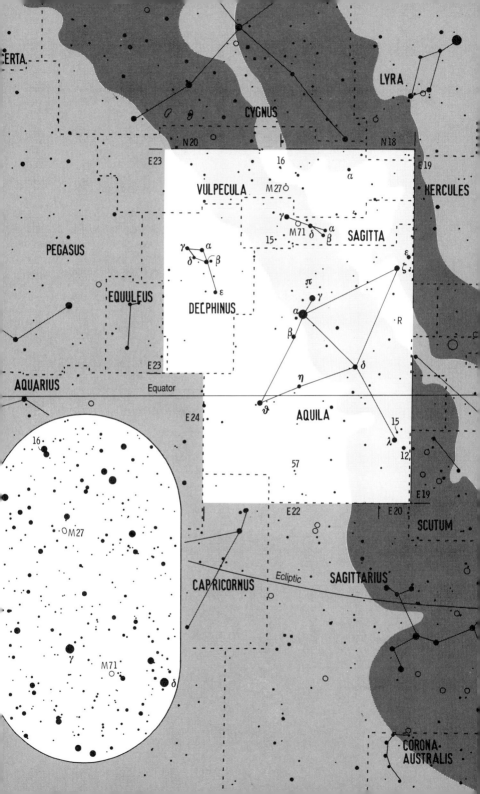

NEBULA		Position		v-Mag.	Size	Shape	Type	Vis.	Dist.	R.A.	Dec.
6809	M 55	Sgr		7	13/□′ 15′	O	XI	GC		18 000 ly	19^{h}40^m.0 −30°.96
6818	.....	Sgr		9½	7	0.4	O	R	PN	6 000	19 44.0 −14.15
6822	.....	Sgr		9	14	12	O	Ir	Glx	2 M	19 45.0 −14.80
6864	M 75	Sgr		9	11	3	O	I	GC	60 000	20 06.1 −21.92
7099	M 30	Cap		7½	11	6	O	V	GC	25 000	21 40.4 −23.18

6809 M 55   Quite large in binoculars, completely resolved in a telescope, irregular outline, dark notch on its southeastern side, difficult to find.

6818 .....   Stellar in binoculars, oval disk in a telescope at high power; it does not look like a ring but the center is a little dim; slightly greenish.

6822 .....   **Barnard's Galaxy**, very close galaxy, very hard to see since there is no detail and no core, darkest sky and lowest power essential.

6864 M 75   Quite distant globular cluster, therefore faint, small, and not resolvable into individual stars, contains an extraordinary bright center.

7099 M 30   Distinct core, elongated envelope, outer portions can be resolved in a telescope, radial chains of stars (see also bottom right of page E1).

STAR		Position		V-Mag.	B−V	Te.	Abs.	Name	Dist.	R.A.	Dec.
44	ϱ^1	Sgr	•	3.9	0.2		1^M	. . . .	122 ly	19^{h}21^m.7 −17°.85
5	α^1	Cap	•	4.3	1.0		−2	Algiedi ⎫ 6′.4	700	20 17.6 −12.51
6	α^2	Cap	•	3.6	0.9		1	Algiedi ⎭ . . .	108	20 18.1 −12.54
9	β	Cap	•	3.0 ✱	0.7		−2	. . . .	320	20 21.0 −14.78
11	ϱ	Cap	•	4.6 ✱	0.5		0	. . . . . .	98,500	20 28.9 −17.82
12	o	Cap	·	5.5 ✱	0.1		1	. . . .	220	20 29.9 −18.58
16	ψ	Cap	•	4.1	0.4		3	. . . .	48	20 46.1 −25.27
18	ω	Cap	•	4.1	1.6		−2	. . . .	600	20 51.8 −26.92
23	ϑ	Cap	•	4.1	0.0		1	. . . .	160	21 05.9 −17.23
32	ι	Cap	•	4.3	0.9		0	. . . .	215	21 22.2 −16.83
34	ζ	Cap	•	3.8	1.0		−2	. . . .	400	21 26.7 −22.41
40	γ	Cap	•	3.7	0.3		1	. . . .	130	21 40.1 −16.66
49	δ	Cap	•	2.8–3.1	0.3		2	Deneb Algedi	38.5	21 47.0 −16.13
12	η	PsA	·	5.4 ✱	−.1		−2	. . . .	1 000	22 00.8 −28.45
17	β	PsA	•	4.3 ✱	0.0		1	. . . .	145	22 31.5 −32.35
18	ε	PsA	•	4.2	−.1		−3	. . . .	700	22 40.7 −27.04
22	γ	PsA	•	4.5 ✱	0.0		0	. . . .	220	22 52.5 −32.88
23	δ	PsA	•	4.2	1.0		1	. . . .	170	22 55.9 −32.54
24	α	PsA	●	1.2	0.1		2	. **Fomalhaut** .	25.2	22 57.6 −29.62

BINARY		Position		V-Mag.		B−V		Te.	Sep.	PA	Vis.
9	β	Cap	•	3.1	6.1	0.8	0.0		205″.2	••	
11	ϱ	Cap	•	4.8	6.6	0.4	1.1		258	⁚	
12	o	Cap	·	5.9	6.7	0.1	0.2		21.9	••	
12	η	PsA	·	5.8	6.8	−.1	0.0		1.8	••	
17	β	PsA	•	4.3	7.8	0.0	0.5		30.3	⁚	
22	γ	PsA	•	4.5	8.0	0.0	0.5		4.1	••	

VARIABLE STAR

49 δ Cap

Period 1.02277 d

Min. 2451200.96

Eclipse 4 hours

The light curve varies slightly.

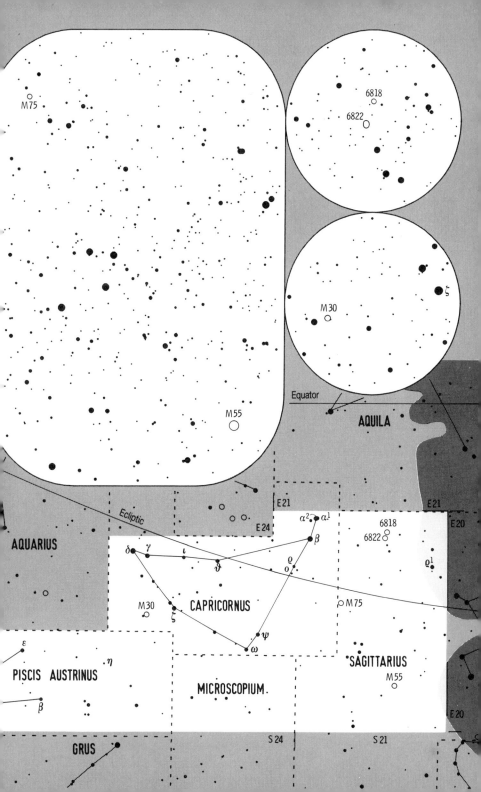

M75

6818
6822

M30

ζ

M55

Equator

AQUILA

E21

E21

E20

α² α¹

β

6818
6822

ϱ¹

E24

δ γ ι

ϑ

ϱ
ο

AQUARIUS

M75

CAPRICORNUS

Ecliptic

M30

ζ

SAGITTARIUS

M55

ε

ψ

ω

η

PISCIS AUSTRINUS

MICROSCOPIUM

β

E20

S24

S21

GRUS

NEBULA	Position	v-Mag.	Size	Shape	Type	Vis.	Dist.	R.A.	Dec.
7078 M 15	Peg ▯	6½	11/□′	10′	○ IV	**GC**	▨	35 000 ly	21ʰ30ᵐ0	12°17
7331	Peg ▯	10	13	9	◖ Sc	**Glx**	▱	60 M	22 37.1	34.42

7078 M 15   Almost stellar in opera glasses, small slightly oval glow in binoculars, resolved in a telescope with the exception of the bright center, relatively easy to find since it lies in a poor region not too far from Enif; this is the best globular cluster in the fall constellations.

7331 .....   Nice spindle, almost edge-on galaxy, oval core within very elongated halo; 30′ to the southwest is Stephan's Quintett, consisting of five mag. 13 galaxies within a 4′ circle, extremely challenging object.

STAR		Position	V-Mag.	B−V	Te.	Abs.	Name	Dist.	R.A.	Dec.
1	ε Equ	▯ ·	5.2	✶ 0.5	↓	1ᴹ	. . . .	200 ly	20ʰ59ᵐ1	4°29
5	γ Equ	▯ •	4.7	0.3	↓	2	. . . .	116	21 10.3	10.13
7	δ Equ	▯ •	4.5	0.5	↓	3	. . . .	61	21 14.5	10.01
8	α Equ	▯ •	3.9	0.5	↓	0	. Kitalphar .	185	21 15.8	5.25
1	Peg	▯ •	4.1	1.1	↓	1	. . . .	155	21 22.1	19.80
8	ε Peg	▯ ●	2.4	1.5	↓	−4	. . . Enif . . .	700	21 44.2	9.88
10	κ Peg	▯ •	4.1	0.4	↓	1	. . . .	115	21 44.6	25.65
24	ι Peg	▯ •	3.8	0.4	↓	3	. . . .	38.5	22 07.0	25.35
29	π Peg	▯ •	4.3	0.5	↓	0	also π² Pegasi	250	22 10.0	33.18
26	ϑ Peg	▯ •	3.5	0.1	↓	1	. . Baham . .	97	22 10.2	6.20
37	Peg	▯ ·	5.5	0.4	↓	2	. . . .	170	22 30.0	4.43
42	ζ Peg	▯ ●	3.4	−.1	↓	−1	. . Homam . .	205	22 41.5	10.83
44	η Peg	▯ ●	2.9	0.8	↓	−1	. . Matar . .	220	22 43.0	30.22
47	λ Peg	▯ •	4.0	1.1	↓	−1	. . . .	380	22 46.5	23.57
46	ξ Peg	▯ •	4.2	0.5	↓	3	. . . .	53	22 46.7	12.17
48	μ Peg	▯ •	3.5	0.9	↓	1	. Sadalbari .	117	22 50.0	24.60
53	β Peg	▯ ●	2.4–2.6	1.7	↓	−2	. . **Scheat** . .	200	23 03.8	28.08
4	β Psc	▯ •	4.5	−.1	↓	−1	. . . .	500	23 03.9	3.82
54	α Peg	▯ ●	2.5	0.0	↓	−1	. . **Markab** . .	140	23 04.8	15.21
6	γ Psc	▯ •	3.7	0.9	↓	1	. . . .	130	23 17.2	3.28
10	ϑ Psc	▯ •	4.3	1.1	↓	1	. . . .	160	23 28.0	6.38
72	Peg	▯ ·	5.0	✶ 1.4	↓	−1	. . . .	500	23 34.0	31.33
17	ι Psc	▯ •	4.1	0.5	↓	3	. . . .	45	23 39.9	5.63
18	λ Psc	▯ •	4.5	0.2	↓	2	. . . .	100	23 42.0	1.78
19	Psc	▯ ·	4.9–5.1	2.5	.	−2	TX Piscium	800	23 46.4	3.49
28	ω Psc	▯ •	4.0	0.4	↓	1	. . . .	106	23 59.3	6.86

BINARY	Position	V-Mag.		B−V		Te.	Sep.	PA	Vis.
1 ε Equ	▯ ·	5.4 ✶ 7.2		0.5 0.5	⇊		10″6	••	▯
		6.0 6.3		0.4 0.5	⇊	′0	0.8	••	●
						2007	0.5	••	▯
						2015	0.2	••	▯
72 Peg	▯	· 5.6 5.9		1.4 1.4	⇊		0.5	••	▯

VARIABLE STAR

53 β Peg ▯ ● irregular
    Extrema 2.3–2.7
19 TX Psc ▯ · irregular
    Extrema 4.8–5.2
    Color orange-red.

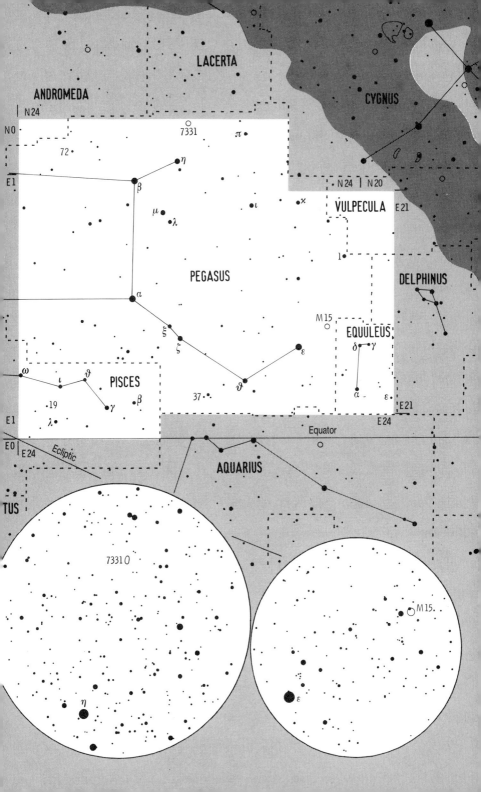

E24 ———— Equator, Ecliptic ———— Fall Constellations

NEBULA	Position	v-Mag.	Size	Shape	Type	Vis.	Dist.	R.A.	Dec.
6981 M72	Aqr ⬡	9½ 12/□'	3'	○ IX	GC	⬡	60000 ly	20ʰ53ᵐ.5	−12°.54
6994 M73	Aqr ⬡	8½ 10	2.5	○ p	OC	⬡	2000	20 59.0	−12.63
7009	Aqr ⬡	8 6	0.6	○ A	PN	⬡	2500	21 04.2	−11.37
7089 M2	Aqr ⬡	6½ 11	10	○ II	GC	⬡	40000	21 33.5	−0.82
7293	Aqr ⬡	7 13	15	○ R	PN	⬡	500	22 29.6	−20.84

6981 M72   The faintest globular cluster in this catalog; it is not resolvable.
6994 M73   Hardly an open cluster, Messier describes it as a group of 3–4 stars.
7009 .....   **Saturn Nebula**, a blue-green ellipse, needs high power; the faint extensions and the mag. 12.8 central star are difficult to observe.
7089 M2   Large bright glow in binoculars, barely resolvable in a telescope.
7293 .....   **Helix Nebula**, the brightest and nearest planetary; the planetary for binoculars, but only at dark sky, needs low power in a telescope; interesting details are visible in the ring through a nebula filter.

STAR		Position	V-Mag.	B−V	Te.	Abs.	Name	Dist.	R.A.	Dec.
2	ε Aqr	⬡ •	3.8	0.0	↓	0ᴹ	....	230 ly	20ʰ47ᵐ.7	−9°.50
12	Aqr	⬡ ·	5.5 ✶	0.7	↓	0	....	500	21 04.1	−5.82
13	ν Aqr	⬡ •	4.5	0.9	↓	1	....	165	21 09.6	−11.37
22	β Aqr	⬡ ●	2.9	0.8	↓	−4	. Sadalsuud .	700	21 31.6	−5.57
34	α Aqr	⬡ ●	3.0	1.0	↓	−4	. Sadalmelik .	800	22 05.8	−0.32
41	Aqr	⬡ ·	5.3 ✶	0.8	↓	1	....	280	22 14.3	−21.07
48	γ Aqr	⬡ •	3.9	−.1	↓	1	. Sadachbia .	150	22 21.7	−1.39
55	ζ Aqr	⬡ •	3.7 ✶	0.4	↓	1	....	105	22 28.8	−0.02
62	η Aqr	⬡ •	4.0	−.1	↓	0	....	180	22 35.4	−0.12
71	τ² Aqr	⬡ •	4.0	1.6	↓	−1	....	400	22 49.6	−13.59
73	λ Aqr	⬡ •	3.7	1.6	↓	−2	....	380	22 52.6	−7.58
76	δ Aqr	⬡ •	3.3	0.1	↓	0	....	170	22 54.6	−15.82
88	Aqr	⬡ •	3.7	1.2	↓	−1	....	240	23 09.4	−21.17
94	Aqr	⬡ ·	5.1 ✶	0.8	↓	3	....	70	23 19.1	−13.46
98	Aqr	⬡ •	4.0	1.1	↓	0	....	160	23 23.0	−20.10
101	Aqr	⬡ ·	4.7 ✶	0.0	↓	0	....	320	23 33.3	−20.91
104	Aqr	⬡ ·	4.8	0.8	↓	−2	....	700	23 41.8	−17.81
R	Aqr	⬡ ·	6.0–10	1.5	↓	−1	....	800	23 43.8	−15.28
107	Aqr	⬡ ·	5.3 ✶	0.3	↓	1	....	210	23 46.0	−18.68

BINARY		Position	V-Mag.		B−V		Te.	Sep.	PA	Vis.
12	Aqr	⬡ ·	5.8	7.4	0.8	0.1	↓↓	2".4	⦂	⬡
41	Aqr	⬡ ·	5.6	7.1	0.9	0.4	↓↓	5.2	•⦂	⬡
55	ζ Aqr	⬡ •	4.3	4.5	0.4	0.5	↓↓'0	2.1	⦂	⬡
			2020 ↗ 1995				2007	2.3	⦂	⬡
							2015	2.6	⦂	⬡
94	Aqr	⬡ ·	5.2	7.4	0.8	0.9	↓↓	12.5	⦂	⬡
101	Aqr	⬡ ·	4.8	7.2	0.0	0.2	↓↓	0.9	•⦂	⬡
107	Aqr	⬡ ·	5.7	6.7	0.3	0.3	↓↓	6.9	•⦂	⬡

VARIABLE STAR

R Aqr ⬡ ·

Period	388 d
Max.	2451333
Min.	Max.+220
Extrema	5.8–12.4

Period increases and decreases every 24 years.

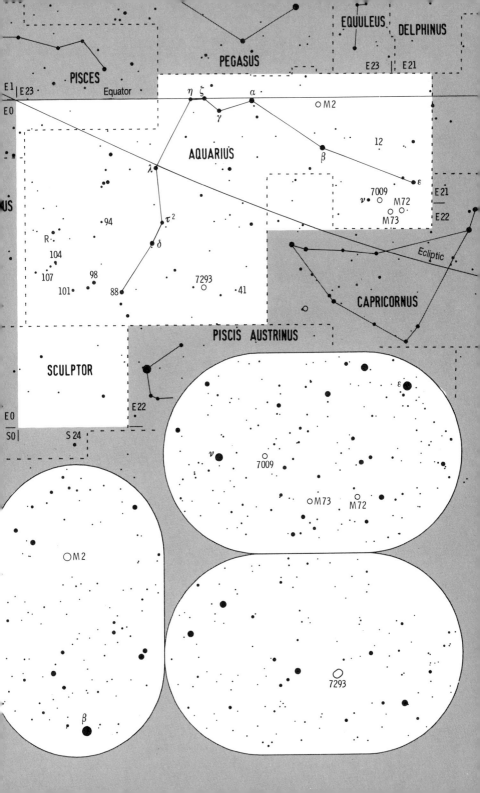

NEBULA	Position	v-Mag.	Size	Shape	Type	Vis.	Dist.	R.A.	Dec.
55	Scl	8	13/□′	25′	❙ Sm	Glx	10 Mly	$0^h15^m.1$	$-39°.22$
104	Tuc	4	11	25	O III	GC	14 000	0 24.1	-72.08
292 SMC	Tuc	2½	13	180	0 Sm	Glx	200 000	0 53	-72.8
362	Tuc	6½	11	10	O III	GC	30 000	1 03.2	-70.86
1291	Eri	9	12	6	0 S0	Glx	40 M	3 17.3	-41.11
1316	For	9	11	3.5	0 S0	Glx	80 M	3 22.7	-37.21
1365	For	10	13	6	0 Sb	Glx	80 M	3 33.6	-36.14

55 ..... Very elongated, bright western part, bright knots, needs low power.
104 ..... **47 Tucanae**, majestic globular cluster, even with unaided eye, outstanding core, huge number of stars in a telescope, $2°$ west of SMC.
292 SMC **Small Magellanic Cloud**, eye-catching under dark sky with unaided eye, nice features in northern portion, low power necessary.
362 ..... Bright distinct center, outer region slightly resolved in a telescope.
1291 ..... Circular central area, elongated halo, hardly any features visible.
1316 ..... **Fornax A**, brightest galaxy of the **Fornax Cluster**, stellar core.
1365 ..... Bar of the barred spiral barely visible, many mag. 11 galaxies nearby.

STAR	Position	V-Mag.	B−V	Te.	Abs.	Name	Dist.	R.A.	Dec.
ε	Phe	3.9	1.0	↓	1^M	....	140 ly	$0^h09^m.4$	$-45°.75$
β	Hyi	2.8	0.6	↓	3	....	24.4	0 25.8	-77.25
κ	Phe	3.9	0.2	↓	2	....	77	0 26.2	-43.68
α	Phe	2.4	1.1	↓	1	.. Ankaa ..	77	0 26.3	-42.31
β	Tuc	3.7 ✶	0.0	↓	1	....	140	0 31.6	-62.96
β	Phe	3.3	0.9	↓	-1	....	230	1 06.1	-46.72
ζ	Phe	3.9–4.4✶	$-.1$	↓	-1	....	290	1 08.4	-55.25
γ	Phe	3.4	1.5	↓	-1	....	240	1 28.4	-43.32
δ	Phe	3.9	1.0	↓	1	....	145	1 31.3	-49.07
α	Eri	0.5	$-.2$	↓	-3	. **Achernar** .	143	1 37.7	-57.24
p	Eri	5.0 ✶	0.9	↓	5	....	26.6	1 39.8	-56.20
χ	Eri	3.7	0.8	↓	2	....	57	1 56.0	-51.61
α	Hyi	2.9	0.3	↓	1	....	71	1 58.8	-61.57
φ	Eri	3.6	$-.1$	↓	0	....	155	2 16.5	-51.51
ι	Eri	4.1	1.0	↓	1	....	145	2 40.7	-39.86
R	Hor	5.7–12	1.3	↓	-1	....	800	2 53.9	-49.89
ϑ	Eri	2.9 ✶	0.1	↓	-1	.. Acamar ..	160	2 58.3	-40.30
f	Eri	4.3 ✶	0.0	↓	1	....	170	3 48.6	-37.62

BINARY	Position	V-Mag.		B−V		Te.	Sep.	PA	Vis.
β	Tuc	4.4	4.5	$-.1$	0.1	↓↓	27″.0		
ζ	Phe	4	6.9	$-.1$	0.5	↓↓	6.6		
p	Eri	5.8	5.8	0.9	0.8	↓↓	′0 11.4		
							2015 11.6		
ϑ	Eri	3.2	4.3	0.2	0.1	↓↓	8.3		
f	Eri	4.8	5.4	0.0	0.0	↓↓	8.2		

VARIABLE STAR

ζ Phe
Period 1.66976 d
Min. 2451201.43

R Hor
Period 405 d
Max. ≈ 2451572

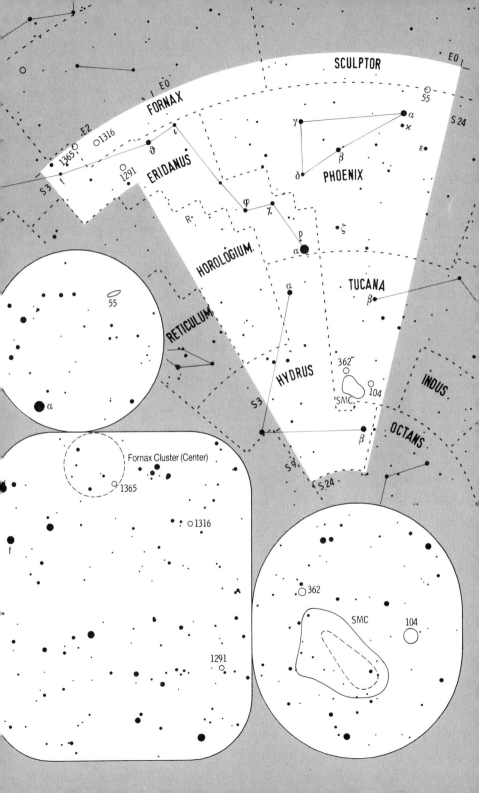

NEBULA	Position	v-Mag.	Size	Shape	Type	Vis.	Dist.	R.A.	Dec.
1851 .. Col	▣	7½ 11/□′	6′	O II	**GC**	⸬	40 000 ly	5ʰ14ᵐ.1	−40°.04
LMC Dor	▣	0 13	420	O Sm	**Glx**	▦	180 000	5 24	−69.8
2070 .. Dor	▣	4½ 11	25	O Em	**DN**	⸬	180 000	5 38.7	−69.10
2516 .. Car	▣	4 11	40	O r	**OC**	⸬	1 200	7 58.3	−60.87

1851 .. Central condensation is well visible in binoculars, hardly resolvable.

LMC **Large Magellanic Cloud**, the brightest and largest nebula, the bar and traces of spiral arms with unaided eye, in binoculars under dark sky past all description, still better in a telescope, many bright knots.

2070 .. **Tarantula Nebula**, fantastic detail, unique in the sky, 5 000 times as luminous as the Orion Nebula, supernova 1987A remnant 20′ southwest.

2516 .. Impressively rich in binoculars, even better in a telescope, rewarding.

STAR		Position	V-Mag.	B−V	Te.	Abs.	Name	Dist.	R.A.	Dec.
β	Ret	▣ ●	3.8	1.1	↓	1ᴹ		100 ly	3ʰ44ᵐ.2	−64°.81
γ	Hyi	▣ ●	3.3	1.6	↓	−1		210	3 47.2	−74.24
δ	Ret	▣ •	4.6	1.6	↓	−2		550	3 58.7	−61.40
α	Hor	▣ ●	3.9	1.1	↓	1		117	4 14.0	−42.29
α	Ret	▣ ●	3.3	0.9	↓	0		165	4 14.4	−62.47
γ	Dor	▣ •	4.3	0.3	↓	3		66	4 16.0	−51.49
ε	Ret	▣ •	4.4	1.1	↓	3		60	4 16.5	−59.30
α	Dor	▣ •	3.3	−.1	↓	0		180	4 34.0	−55.05
ι	Pic	▣ ·	5.2 ✶	0.4	↓	2		120	4 50.9	−53.46
β	Dor	▣ •	3.4–4.1	0.6	↓	−5		1 200	5 33.6	−62.49
β	Pic	▣ •	3.9	0.2	↓	2		63	5 47.3	−51.07
γ	Pic	▣ •	4.5	1.1	↓	1		175	5 49.8	−56.17
η	Col	▣ •	4.0	1.1	↓	−2		500	5 59.1	−42.82
α	Car	▣ ●	−0.7	0.2	↓	−6 .	**Canopus** .	310	6 24.0	−52.70
ν	Pup	▣ ●	3.2	−.1	↓	−2		410	6 37.8	−43.20
α	Pic	▣ ●	3.2	0.2	↓	1		88	6 48.2	−61.94
τ	Pup	▣ ●	2.9	1.2	↓	−1		180	6 49.9	−50.61
γ	Vol	▣ •	3.6 ✶	0.9	↓	1		135	7 08.8	−70.50
δ	Vol	▣ •	4.0	0.8	↓	−3		700	7 16.8	−67.96
ζ	Vol	▣ •	3.9	1.0	↓	1		132	7 41.8	−72.61
χ	Car	▣ •	3.5	−.2	↓	−2		400	7 56.8	−52.98
κ	Vol	▣ ·	4.7 ✶	−.1	↓	−1		400	8 19.9	−71.51
ε	Car	▣ ●	1.9	1.2	↓	−5 . .	**Avoir** . .	600	8 22.5	−59.51
β	Vol	▣ •	3.8	1.1	↓	1		108	8 25.7	−66.14
c	Car	▣ •	3.8	−.1	↓	−1		320	8 55.0	−60.64
α	Vol	▣ •	4.0	0.1	↓	1		124	9 02.4	−66.40

BINARY		Position	V-Mag.		B−V		Te.	Sep.	PA	Vis.
ι	Pic	▣ ·	5.6	6.4	0.3	0.5	⇈	12″.5	•⸰	⸫
γ	Vol	▣ •	3.8	5.7	1.0	0.5	⇊	14.2	•·	⸫
κ	Vol	▣ •	5.3	5.6	−.1	−.1	⇊	64.8	•⸰	⸬

VARIABLE STAR

β	Dor	▣ ●	⟋‾⟍
	Period		9.8425 d
	Max.		2451200.5

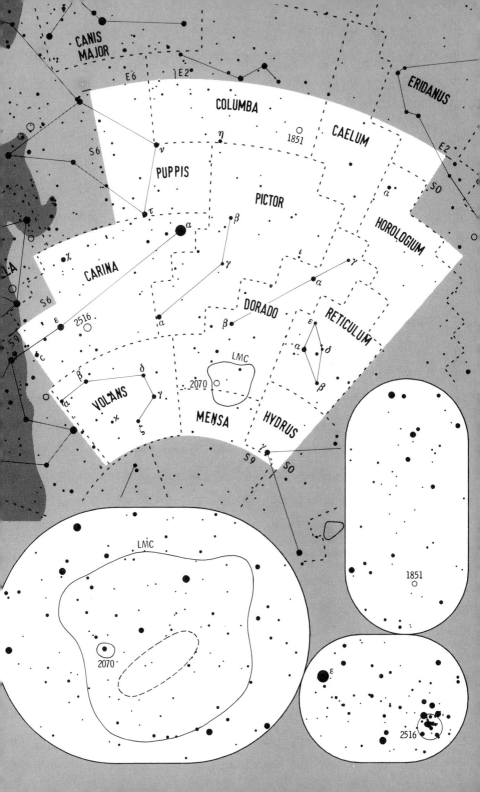

NEBULA Position v-Mag. Size Shape Type Vis. Dist. R.A. Dec.

						Shape	Type	Vis.	Dist.	R.A.	Dec.
2451	..	Pup		3½	11/□′	40′	0 p	OC		1400 ly	7ʰ45ᵐ.4 −37°.97
2477	..	Pup		6	12	25	0 r	OC		4500	7 52.3 −38.53
2547	..	Vel		5	11	20	0 m n	OC		1500	8 10.7 −49.27
2546	..	Pup		6½	13	35	0 m	OC		3000	8 12.4 −37.63
IC 2391		Vel		3	11	40	0 p	OC		480	8 40.2 −53.07
3132	..	Vel		9	9	1.0	0 R	PN		5000	10 07.0 −40.44
3201	..	Vel		7	12	12	0 X	GC		16000	10 17.6 −46.41

2451 .. Well resolved in binoculars, few stars, but very bright and colored ones.
2477 .. Enormous number of stars, nicely resolved in a telescope, chain of stars.
2547 .. Bright open cluster, best in binoculars, a mag. 6.5 star near the center.
2546 .. Very elongated cluster in binoculars, inconspicuous in a telescope.
IC 2391 .. **Omikron (*o*) Velorum Cluster**, for unaided eye to binoculars, sparse.
3132 .. Oval disk with a magnitude 10.1 star slightly off center in a telescope.
3201 .. Irregular stellar condensations; it is just resolvable in a telescope.

STAR Position V-Mag. B−V Te. Abs. Name Dist. R.A. Dec.

			V-Mag.	B−V	Te.	Abs.	Name	Dist.	R.A.	Dec.
L²	Pup	•	4.0–5.0	1.5		0ᴹ	. . . .	185 ly	7ʰ13ᵐ.5 −44°.64
π	Pup	•	2.7	1.6		−5	} Sep. 26′	1000	7 17.1 −37.10
y υ	Pup	•	4.1	−.1		−3	}	1000	7 18.4 −36.74
σ	Pup	•	3.3	1.5		−1	. . . .	185	7 29.2 −43.30
c	Pup	•	3.6	1.7		−5	. in NGC 2451 .	1400	7 45.3 −37.97
a	Pup	•	3.7	1.0		−1	. . . .	340	7 52.2 −40.58
ζ	Pup	●	2.2	−.3		−6	. . . .	1400	8 03.6 −40.00
γ	Vel	●	1.5–1.7	−.2		−6	Suhail Al Muhlif	900	8 09.5 −47.34
o	Vel	•	3.6	−.2		−2	. . in IC 2391 . .	480	8 40.3 −52.92
b	Vel	•	3.8	0.7		−7	. . . .	4000	8 40.6 −46.65
d	Vel	•	4.0	0.9		0	. . . .	225,85	8 44.4 −42.65
δ	Vel	●	1.9	0.0		0	. . . .	80	8 44.7 −54.71
a	Vel	•	3.9	0.0		−4	. . . .	1200	8 46.0 −46.04
c	Vel	•	3.8	1.2		−1	. . . .	310	9 04.2 −47.10
λ	Vel	●	2.2	1.7		−4	Suhail Al Wazn	550	9 08.0 −43.43
κ	Vel	●	2.5	−.1		−4	. . . .	550	9 22.1 −55.01
ψ	Vel	•	3.6	0.4		2	. . . .	60	9 30.7 −40.47
φ	Vel	●	3.5	−.1		−6	. . . .	2000	9 56.9 −54.57
q	Vel	•	3.9	0.1		1	. . . .	102	10 14.7 −42.12
p	Vel	•	3.8	0.3		2	. . . .	87	10 37.3 −48.23
x	Vel	•	4.1	0.8		−3	. . . .	800	10 39.3 −55.60
μ	Vel	•	2.7	0.9		0	. . . .	115	10 46.8 −49.42

BINARY Position V-Mag. B−V Te. Sep. PA Vis.

			V-Mag.		B−V		Te.	Sep.	PA	Vis.
y υ	Pup	•	4.7	5.1	−.1	−.2		240″.0		
γ	Vel	●	2	4.2	−.2	−.2		41.2		
d	Vel	•	4.1	7.2	0.9	0.7		238		
x	Vel	•	4.3	6.2	1.0	−.1		51.8		

VARIABLE STAR

L²	Pup		•	semireg.
Period				140 d
γ	Vel		●	irregular
Period				≈ 2 min.

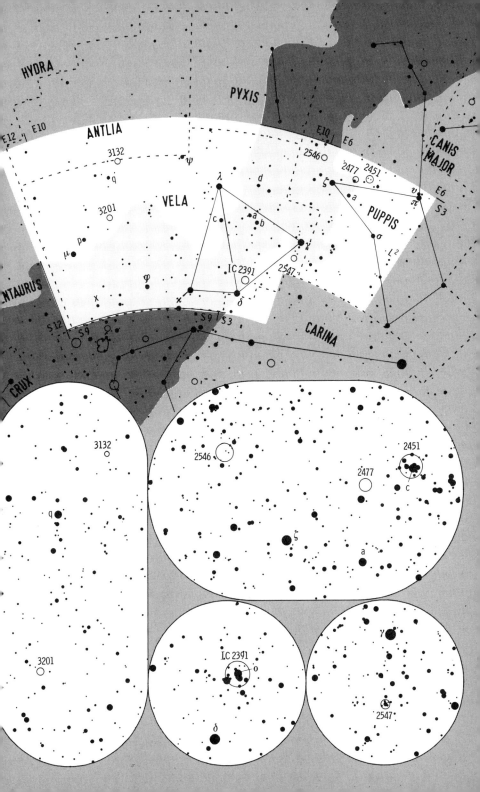

NEBULA Position v-Mag. Size Shape Type Vis. Dist. R.A. Dec.

Name		Pos	v-Mag.	Size	Shape	Type	Vis.	Dist.	R.A.	Dec.
2808 ..	Car	♉	6½ 11/□′	8′	○ I	GC	🌑	30 000 ly	9ʰ12ᵐ0	−64°86
3114 ..	Car	♉	4½ 11	30	○ r	OC	⠿	3 000	10 02.7	−60.12
3293 ..	Car	◦	5 9	6	○ m n	OC	🌑	6 000	10 35.8	−58.23
IC 2602	Car	◦	2 10	60	○ m	OC	⠿	460	10 43.2	−64.40
3372 ..	Car	◦	3 13	100	○ Em	DN	⣿	6 000	10 43.8	−59.87
3532 ..	Car	◦	3½ 12	60	○ r	OC	⠿	1 200	11 06.4	−58.67

2808 .. Resolved in a telescope, shape I = extreme central condensation.
3114 .. Impressively rich open cluster, a rewarding object in every scope.
3293 .. Small glow in binoculars, resolved in a telescope, takes high power.
IC 2602 **Southern Pleiades**, similar to the Pleiades, only a little fainter.
3372 .. **Eta (η) Carinae Nebula**, conspicuous already with unaided eye, full of features in binoculars, even better in a telescope at low power.
3532 .. Extremly rich, elongated open cluster, impressive in every scope.

STAR Position V-Mag. B−V Te. Abs. Name Dist. R.A. Dec.

Star			V-Mag.	B−V	Te.	Abs.	Name	Dist.	R.A.	Dec.
α	Cha	• •	4.1	0.4	↓	3ᴹ	. . . .	63 ly	8ʰ18ᵐ5	−76°92
b¹	Car	• •	4.7 ☆	−.2	↓	−2	. . . .	600	8 57.0	−59.23
a	Car	• •	3.4	−.2	↓	−2	. . . .	430	9 11.0	−58.97
i	Car	♉ •	4.0	−.2	↓	−2	. . . .	500	9 11.3	−62.32
β	Car	• ●	1.7	0.1	↓	−1	Miaplacidus	112	9 13.2	−69.72
ι	Car	• ●	2.2	0.2	↓	−4	. . . .	650	9 17.1	−59.28
N	Vel	• ●	3.2	1.5	↓	−1	. . . .	230	9 31.2	−57.03
R	Car	• ·	4.5–9.2	1.2	↓	−1	. . . .	400	9 32.2	−62.79
l	Car	•◦	3.3–4.1	1.0	↓	−5	. ZZ Carinae .	1 400	9 45.2	−62.51
υ	Car	• ●	2.9 ☆	0.3	↓	−6	. . . .	1 500	9 47.1	−65.07
ω	Car	• ●	3.3	−.1	↓	−2	. . . .	370	10 13.7	−70.04
q	Car	◦	3.4	1.5	↓	−3	. . . .	700	10 17.1	−61.33
I	Car	• •	3.9 ☆	0.3	↓	0	. . . .	53, 430	10 24.4	−74.02
s	Car	◦ •	3.8	0.3	↓	−4	. . . .	1 000	10 27.9	−58.74
p	Car	◦ •	3.3	−.1	↓	−3	. . . .	520	10 32.0	−61.69
γ	Cha	• •	4.1	1.6	↓	−1	. . . .	400	10 35.5	−78.61
ϑ	Car	◦ ●	2.7	−.2	↓	−3	. in IC 2602 .	460	10 43.0	−64.39
δ	Cha	• •	4.1 ☆	0.0	↓	−1	. . . .	370	10 45.6	−80.52
u	Car	◦ •	3.7 ☆	0.8	↓	−2	. . . . 96, 1 500		10 53.5	−58.86
x	Car	◦ •	3.9	1.2	↓	−8	near NGC 3532 5 000		11 08.6	−58.97
ε	Cha	• •	4.7 ☆	0.0	↓	−1	. . . .	380	11 59.7	−78.22

BINARY Position V-Mag. B−V Te. Sep. PA Vis.

Binary			V-Mag.	B−V	Te.	Sep.	PA	Vis.
b¹	Car	• •	4.9 6.8	−.2 −.1	↓↓	40″1	••	⚬
υ	Car	• ●	3.0 6.0	0.3 0.1	↓↓	5.0	•⚬	•
I	Car	• •	4.0 6.2	0.4 0.1	↓↓	232	•	🌑
δ	Cha	• •	4.5 5.5	−.2 1.0	↓↓	265.1	•	🌑
u	Car	◦ •	3.8 6.3	0.9 −.1	↓↓	159	••	🌑
ε	Cha	• •	4.9 6.6	−.1 0.2	↓↓	134.0	••	⠿

VARIABLE STAR

R	Car	• ·		〰
	Period		307 d	
	Max.		2451215	
l ZZ	Car	◦ •		〰
	Period		35.54 d	
	Max.		2451222	

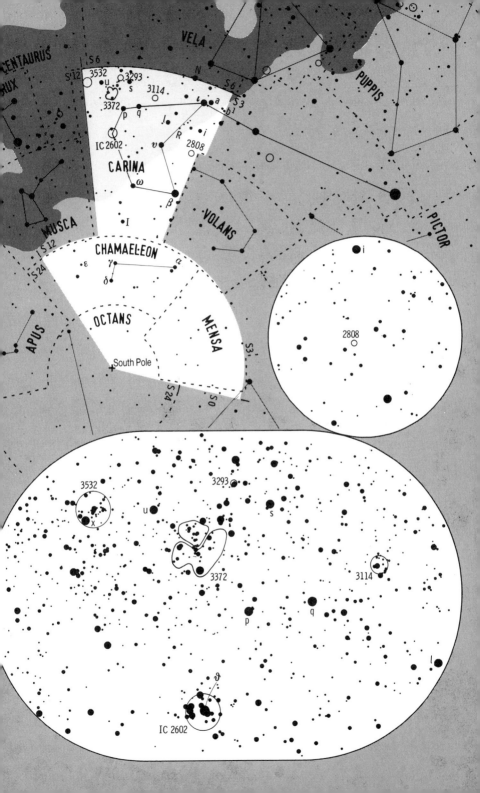

NEBULA	Position		v-Mag.	Size	Shape	Type	Vis.	Dist.	R.A.	Dec.
3766	Cen	⚬	5	10/□′	12′	⚬ m	OC	⊠	5000 ly	11ʰ36ᵐ.1 −61°.62
Coalsack	Cru	♂	(3) (14)	360		0 Dark Neb.		⊠	2000	12 52 −63.3
4755	Cru	♂	4½	9	10	⚬ m	OC	⊞	5000	12 53.6 −60.33

3766 ..... Resolved in binoculars, wonderful in a telescope, interesting shape; some of the brightest stars shine conspicuously in a yellow color.

Coalsack Most spectacular dark nebula for the unaided eye, rich detail in binoculars, northern edge round, southern edge irregular and fuzzy.

4755 ..... **Jewel Box, Kappa (κ) Crucis Cluster**, resolved in binoculars, impressive in a telescope, arrow shaped with a magnitude 5.8 star at the arrow head and the magnitude 6.0 star κ Crucis at the southern end of the arrow, in between lies a yellow star, takes high power.

STAR	Position			V-Mag.	B−V	Te.	Abs.	Name	Dist.	R.A.	Dec.
π	Cen	•	•	3.9	−.2	↧	−1ᴹ		330 ly	11ʰ21ᵐ.0 −54°.49
o	Cen	•	•	4.3–4.5⚹	0.7	↧	−7		5000	11 31.8 −59.48
λ	Cen	⚬	•	3.1	0.0	↧	−2		420	11 35.8 −63.02
λ	Mus	•	•	3.6	0.2	↧	1		130	11 45.6 −66.73
δ	Cen	•	●	2.4 ⚹	−.1	↧	−3		380	12 08.3 −50.71
ϱ	Cen	•	•	4.0	−.2	↧	−1		330	12 11.7 −52.37
δ	Cru	♂	●	2.8	−.2	↧	−3		370	12 15.1 −58.75
ε	Mus	•	•	4.0–4.3	1.6	↟	−1		300	12 17.6 −67.96
ζ	Cru	♂	•	4.1	−.2	↧	−1		370	12 18.4 −64.00
ε	Cru	♂	•	3.5–3.6	1.4	↟	−1		230	12 21.4 −60.40
α	Cru	♂	●	0.7 ⚹	−.2	↧	−4 .. **Acrux** ..		340	12 26.6 −63.10
σ	Cen	•	•	3.9	−.2	↧	−2		450	12 28.0 −50.23
γ	Cru	♂	●	1.6 ⚹	1.5	↟	−1 .. **Gacrux** .	88,400	12 31.2 −57.11
γ	Mus	•	•	3.8	−.2	↧	−1		320	12 32.5 −72.13
α	Mus	•	●	2.7	−.2	↧	−2		300	12 37.2 −69.14
τ	Cen	•	•	3.9	0.0	↧	1 ⎱ Sep. 45′ ••		130	12 37.7 −48.54
γ	Cen	•	●	2.2	0.0	↧	−1 ⎰		130	12 41.5 −48.96
β	Mus	•	•	3.0 ⚹	−.2	↧	−2		320	12 46.3 −68.11
β	Cru	♂	●	1.3	−.2	↧	−4 .. **Mimosa** ..		340	12 47.7 −59.69
μ	Cru	•	•	3.7 ⚹	−.2	↧	−2		370	12 54.6 −57.18
δ	Mus	•	•	3.6	1.2	↟	1		90	13 02.3 −71.55

BINARY	Position			V-Mag.		B−V		Te.	Sep.	PA	Vis.
o	Cen	•	•	5	5.1	1.0	0.4	↥↧	265″.7	⁚	⊠
δ	Cen	•	●	2.6	4.5	−.1	−.2	↧↧	269.0	•⁚	⊠
				″	6.4	″	0.0	↧	216.7	•⁚	⊞
α	Cru	♂	●	0.8⚹	4.8	−.2	−.1	↧↧	90.0	⁚•	⚬
				1.3	1.7	−.2	−.2	↧↧	4.0	•⁚	⚬
γ	Cru	♂	●	1.6	6.4	1.6	0.2	↟↧	128	⁚•	⚬
β	Mus	•	•	3.6	4.0	−.2	−.2	↧↧	1.1	•⁚	⚬
μ	Cru	•	•	4.0	5.1	−.2	−.1	↧↧	34.8	⁚•	⚬

VARIABLE STAR

o Cen ▣ • semireg.
Period ≈ 200 d ?
Binary star mag.
5.0–5.3 and 5.1.

ε Mus ▣ • semireg.
Period ≈ 40 d

ε Cru ♂ • irreg. ?
Extrema 3.4–4.0

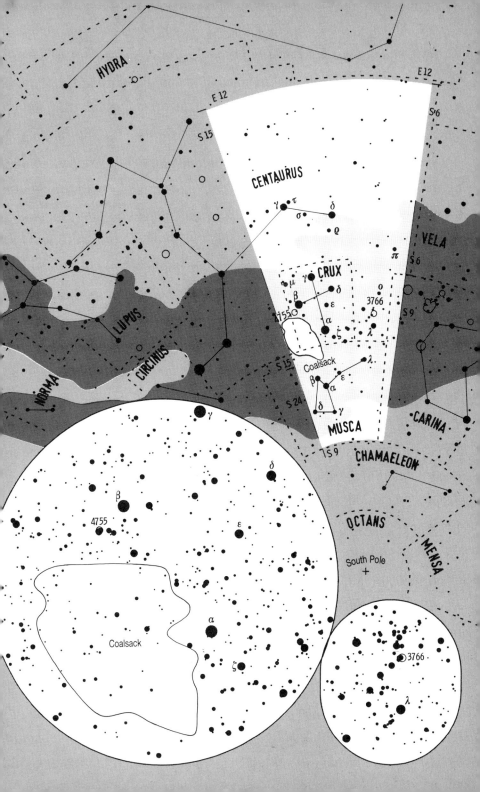

NEBULA Position v-Mag. Size Shape Type Vis. Dist. R.A. Dec.

					Size	Shape	Type	Vis.	Dist.	R.A.	Dec.
5128	..	Cen	♂	7½ 12/□′	12′	0 S0	Glx	⟦∷⟧	20 M ly	13ʰ25ᵐ.5	−43°.02
5139	..	Cen	♂	4 11	30	0 VIII	GC	⟦∷⟧	16 000	13 26.8	−47.48
5460	..	Cen	⊙	6 13	30	0 p	OC	⟦∷⟧	2 500	14 07.6	−48.32
5822	..	Lup	♂	6½ 14	40	0 r	OC	⟦∷⟧	2 500	15 05.2	−54.33

5128 .. **Centaurus A**, a dark dust band with features, a remarkable galaxy.
5139 .. **Omega (ω) Centauri**, largest, brightest, and most luminous globular cluster, easy with unaided eye, bright elliptical glow in binoculars, tremendous richness of stars in a telescope, unsurpassed, fascinating.
5460 .. Resolved in binoculars, widely scattered in a telescope at low power.
5822 .. Easy object for binoculars under dark sky, ample stars in a telescope.

STAR Position V-Mag. B−V Te. Abs. Name Dist. R.A. Dec.

Star			V-Mag.	B−V	Te.	Abs.	Name	Dist.	R.A.	Dec.
ι	Cen	•	2.8	0.1	⌊	1ᴹ		58 ly	13ʰ20ᵐ.6	−36°.71
J	Cen	•	4.3 ✶	−.1	⌊	−1		360	13 22.6	−60.99
d	Cen	•	3.9	1.2	⌊	−4		1 000	13 31.0	−39.41
ε	Cen	•	2.3	−.2	⌊	−3		380	13 39.9	−53.47
Q	Cen	•	5.0 ✶	0.0	⌊	0		270	13 41.7	−54.56
ν	Cen	♂	3.4	−.2	⌊	−3 } Sep. 47′		500	13 49.5	−41.69
μ	Cen	♂	3.0–3.5	−.2	⌊	−3 }		500	13 49.6	−42.47
ζ	Cen	♂	2.5	−.2	⌊	−3		400	13 55.5	−47.29
φ	Cen	♂	3.8	−.2	⌊	−2		500	13 58.3	−42.10
υ¹	Cen	♂	3.9	−.2	⌊	−2		400	13 58.7	−44.80
β	Cen	•	0.6	−.2	⌊	−5	**Hadar, Agena**	520	14 03.8	−60.37
5 ϑ	Cen	•	2.1	1.0	⌊	1		62	14 06.7	−36.37
R	Cen	•	5.8–10	1.9	⌊	−3		2 000	14 16.6	−59.91
η	Cen	•	2.3	−.2	⌊	−3		300	14 35.5	−42.16
α	Cen	•	−0.3 ✶	0.7	⌊	4	**Rigil Kentaurus**, ⌊Toliman⌋	4.40	14 39.6	−60.83
b	Cen	•	4.0	−.2	⌊	−1		300	14 42.0	−37.79
α	Cir	•	3.2	0.3	⌊	2		53	14 42.5	−64.98
κ	Cen	•	3.1	−.2	⌊	−3		500	14 59.2	−42.10
δ	Cir	•	4.6 ✶	−.1	⌊	−6		3 000	15 16.8	−60.94
β	Cir	•	4.1	0.1	⌊	2		98	15 17.5	−58.80
γ	Cir	•	4.5	0.2	⌊	−1		450	15 23.4	−59.32

BINARY Position V-Mag. B−V Te. Sep. PA Vis.

Binary			V-Mag.	B−V	Te.	Sep.	PA	Vis.
J	Cen	•	4.5 6.2	−.1 0.0	⌊⌊	60″.6		⟦∷⟧
Q	Cen	•	5.3 6.6	−.1 0.1	⌊⌊	5.5		⟦·⟧
α	Cen	•	0.0 1.4	0.7 0.9	⌊⌊ ′0	14.1		⟦·⟧

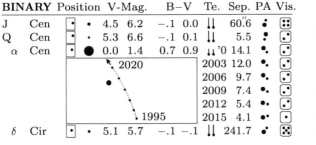

	2003	12.0	⟦·⟧
	2006	9.7	⟦·⟧
	2009	7.4	⟦·⟧
	2012	5.4	⟦·⟧
	2015	4.1	⟦·⟧

| δ | Cir | • | 5.1 5.7 | −.1 −.1 | ⌊⌊ | 241.7 | | ⟦∷⟧ |

VARIABLE STAR

μ Cen ♂ • irregular
R Cen • ⟦∿⟧
Period 550 d
Max. 2451355

Proxima Centauri
Closest star, 4.22 ly,
11ᵐ companion of and
2° south of Toliman.

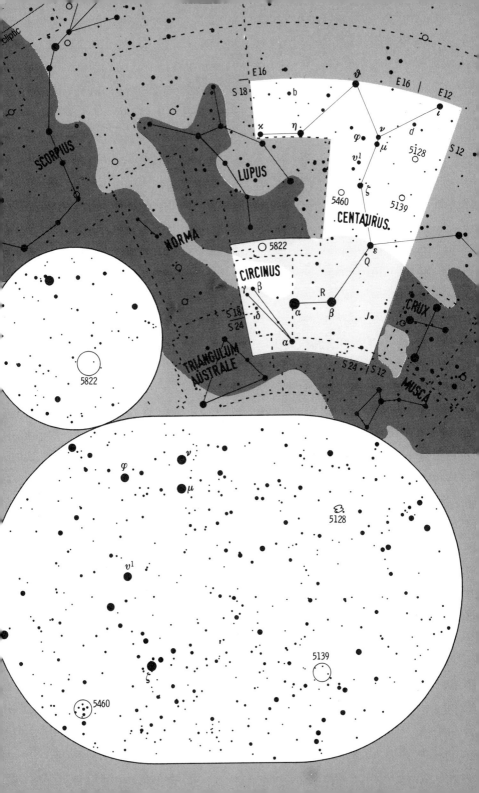

NEBULA	Position		v-Mag.	Size	Shape	Type	Vis.	Dist.	R.A.	Dec.
5986 ..	Lup	⬡	7½	11/□' 5'	O VII	GC	⬚	35000 ly	15h46m.1	−37°.79
6067 ..	Nor	⬡	6	11 15	O r	OC	⬚	6000	16 13.2	−54.22
6087 ..	Nor	⬡	6	11 15	O p	OC	⬚	3000	16 18.9	−57.90
6397 ..	Ara	⬡	6	12 20	O IX	GC	⬚	7000	17 40.7	−53.67

5986 .. Uniform round glow in binoculars, outer region resolved in a telescope.
6067 .. Bright open cluster suited for binoculars; it is very rich in a telescope.
6087 .. Nicely resolved in binoculars, sparse; it needs low power in a telescope.
6397 .. Triangular in binoculars, well resolved in a telescope, chains of stars.

STAR	Position		V-Mag.	B−V	Te.	Abs.	Name	Dist.	R.A.	Dec.
ι	Lup	•	3.5	−.2		-2^M	....	330 ly	14h19m.4	−46°.06
α	Lup	•	2.3	−.2		−4	....	500	14 41.9	−47.39
β	Lup	•	2.7	−.2		−3	....	500	14 58.5	−43.13
π	Lup	•	3.9	☆ −.1		−2	....	500	15 05.1	−47.05
κ	Lup	•	3.7	☆ 0.0		0	....	185	15 11.9	−48.74
ζ	Lup	•	3.4	☆ 0.9		1	....	115	15 12.3	−52.10
μ	Lup	•	4.1	☆ 0.0		−1	. (see below) .	270	15 18.5	−47.88
δ	Lup	•	3.2	−.2		−3	....	600	15 21.4	−40.65
φ^1	Lup	•	3.6	1.5		−1	....	310	15 21.8	−36.26
ε	Lup	•	3.4	−.2		−3	....	500	15 22.7	−44.69
γ	Lup	•	2.8	−.2		−4	....	600	15 35.1	−41.17
d	Lup	•	4.5	☆ −.2		−1	....	450	15 35.9	−44.96
η	Lup	•	3.4	☆ −.2		−3	....	500	16 00.1	−38.40
γ^2	Nor	•	4.0	1.1		1	....	128	16 19.8	−50.16
ε	Nor	•	4.4	☆ −.1		−1	....	440	16 27.2	−47.55
η	Ara	•	3.8	1.6		−1	....	310	16 49.8	−59.04
ζ	Ara	•	3.1	1.6		−3	....	550	16 58.6	−55.99
ε^1	Ara	•	4.1	1.4		−1	....	300	16 59.6	−53.16
β	Ara	•	2.8	1.5		−3	} Sep. 51'	550	17 25.3	−55.53
γ	Ara	•	3.3	−.1		−5	}	1200	17 25.4	−56.38
δ	Ara	•	3.6	−.1		0	....	185	17 31.1	−60.68
α	Ara	•	2.9	−.1		−1	....	240	17 31.8	−49.88
ϑ	Ara	•	3.7	−.1		−4	....	1200	18 06.6	−50.09

BINARY	Position		V-Mag.		B−V		Te.	Sep.	PA	Vis.
π	Lup	•	4.6	4.7	−.1	−.1	‖	1".7	••	[•]
κ	Lup	•	3.9	5.7	0.0	0.2	‖	26.4	•?	[•]
ζ·	Lup	•	3.4	6.7	0.9	0.5	‖	71.7	••	[•]
μ	Lup	•	4.2☆	7.0	−.1	0.5	‖	240.5	••	[⬚]
			4.3☆	6.8	−.1	0.1	‖	23.2	••	[•]
			5.0	5.1	−.1	−.1	‖	1.0	•'	[•]
d	Lup	•	4.7	6.7	−.2	−.1	‖	2.1	••	[•]
η	Lup	•	3.4	7.8	−.2	0.3	‖	15.0	••	[•]
ε	Nor	•	4.5	7.2	−.1	0.0	‖	22.8	••	[•]

Comment on μ Lupi

It is an interesting multiple star, visible as a binary (4.2, 7.0) in binoculars or in a finder, triple star (4.3, 6.8, 7.0) in a guide scope, and quadruple star (5.0, 5.1, 6.8, 7.0) in a telescope.

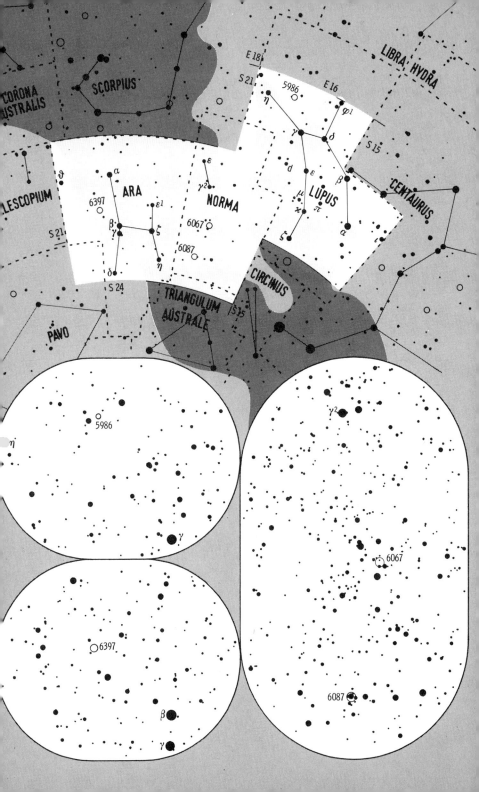

NEBULA		Position	v-Mag.	Size		Shape	Type	Vis.	Dist.	R.A.	Dec.
6124 ..	Sco	♂	6	12/□′	25′	O r	OC	⊠	2 000 ly	16ʰ25.ᵐ6	−40.°67
6231 ..	Sco	♂	3½	9	15	O p n	OC	⊞	5 000	16 54.0	−41.80
6388 ..	Sco	◦	7	10	4	O III	GC	⊡	40 000	17 36.3	−44.73
6541 ..	CrA	◦	6½	10	7	O III	GC	⊠	22 000	18 08.0	−43.71
6723 ..	Sgr	◦	7½	11	6	O VII	GC	⊡	30 000	18 59.6	−36.63

6124 .. Partially resolved in binoculars, shows many faint stars in a telescope.
6231 .. One of the brightest clusters, distinct with unaided eye; a few bright stars stand out in binoculars; some more become visible in a telescope.
6388 .. Easily visible as a nebula, distinct bright core, not resolvable into stars.
6541 .. Oval and asymmetric in binoculars, partially resolved in a telescope.
6723 .. Oval glow in binoculars, resolved in a telescope, central area not much brighter; 30′ to the south is the faint reflection nebula NGC 6726, 6727.

STAR		Position		V-Mag.	B−V	Te.	Abs.	Name	Dist.	R.A.	Dec.
μ^1	Sco	♂	•	2.9–3.2	−.1	↓	−3ᴹ	} Sep. 5.′8 ••	550 ly	16ʰ51.ᵐ9	−38.°05
μ^2	Sco	♂	•	3.6	−.2	↓	−3	}	550	16 52.3	−38.02
ζ^2	Sco	♂	•	3.6	1.4	↓	0	. . . .	150	16 54.6	−42.36
η	Sco	•	•	3.3	0.4	↓	2	. . . .	71	17 12.2	−43.24
34 v	Sco	•	•	2.7	−.2	↓	−4	. . Lesath . .	600	17 30.8	−37.30
35 λ	Sco	•	●	1.6	−.2	↓	−5	. . Shaula . .	600	17 33.6	−37.10
ϑ	Sco	◦	●	1.9	0.4	↓	−3	. . Sargas . .	270	17 37.3	−43.00
κ	Sco	•	●	2.4	−.2	↓	−3	. . . .	450	17 42.5	−39.03
ι^1	Sco	•	●	3.0	0.5	↓	−6	. . . .	2 000	17 47.6	−40.13
G	Sco	•	●	3.2	1.2	↓	0	. . . .	127	17 49.9	−37.04
η	Sgr	•	●	3.1 ✭	1.6	↓	0	. . . .	150	18 17.6	−36.76
α	Tel	•	●	3.5	−.2	↓	−1	. . . .	260	18 27.0	−45.97
ζ	Tel	•	●	4.1	1.0	↓	1	. . . .	130	18 28.8	−49.07
κ	CrA	•	·	5.2 ✭	0.0	↓	−2	. . . .	800	18 33.4	−38.72
ε	CrA	◦	·	4.7–5.0	0.4	↓	2	. . . .	97	18 58.7	−37.11
ζ	CrA	•	·	4.7	0.0	↓	1	. . . .	180	19 03.1	−42.10
γ	CrA	◦	·	4.2 ✭	0.5	↓	3	. . . .	58	19 06.4	−37.06
δ	CrA	•	·	4.6	1.1	↓	1	. . . .	175	19 08.3	−40.50
α	CrA	◦	·	4.1	0.0	↓	1	. . . .	130	19 09.5	−37.90
β	CrA	•	·	4.1	1.2	↓	−2	. . . .	500	19 10.0	−39.34
β^1	Sgr	•	·	3.9 ✭	−.1	↓	−1	. . . .	380	19 22.6	−44.46
α	Sgr	•	·	4.0	−.1	↓	0	. . . .	175	19 23.9	−40.62

BINARY		Position		V-Mag.		B−V	Te.	Sep.	PA	Vis.
η	Sgr	•	●	3.1	7.8	1.6 0.8	↓↓	3.″6	•·	⊙
κ	CrA	•	·	5.6	6.3	−.1 0.0	↓↓	21.4	⦂	⊙
γ	CrA	◦	•	4.9	5.1	0.5 0.5	↓↓ ′0	1.3	•·	⊙
					2020		2007	1.3	⦂	⊙
			1995 •				2015	1.4	⦂	⊙
β^1	Sgr	•	•	4.0	7.1	−.1 0.3	↓↓	28.3	•·	⊙

VARIABLE STAR

μ^1 Sco ♂ • ⌣‾⌣
Period 1.44627 d
Min. 2451200.29

ε CrA ◦ · ‾⌣‾
Period 0.591436 d
Min. 2451200.45

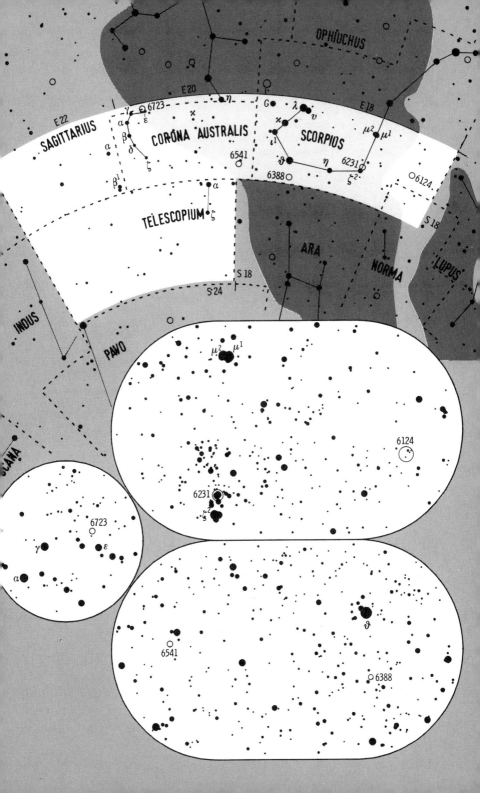

NEBULA Position v-Mag. Size Shape Type Vis. Dist. R.A. Dec.

NEBULA		Position	v-Mag.	Size	Shape	Type	Vis.	Dist.	R.A.	Dec.
6752	.. Pav	⊙•	5½	11/□'	15'	○ VI	GC	13000 ly	19ʰ10ᵐ.9	−59°.98

6752 .. Fourth in brightness among globular clusters, near the limit of the unaided eye, bright nebula in binoculars, well resolved in a telescope, interesting features, remarkable chains of stars, spider-like, rewarding.

STAR Position V-Mag. B−V Te. Abs. Name Dist. R.A. Dec.

STAR		Position	V-Mag.	B−V	Te.	Abs.	Name	Dist.	R.A.	Dec.
δ	Oct	•	4.3	1.3	↓	0^M		280 ly	14ʰ26ᵐ.9	−83°.67
α	Aps	•	3.8	1.4	↓	−2		410	14 47.9	−79.04
γ	TrA	●	2.9	0.0	│	−1		185	15 18.9	−68.68
ε	TrA	•	4.1	1.2	↓	0		215	15 36.7	−66.32
β	TrA	●	2.8	0.3	│	2		40	15 55.1	−63.43
δ	TrA	•	3.9	1.1	↓	−3		600	16 15.4	−63.69
δ	Aps	•	4.2 ☆	1.6	↓	−3 } Sep. 40' ••		700	16 20.4	−78.69
γ	Aps	•	3.9	0.9	↓	0		155	16 33.5	−78.90
β	Aps	•	4.2	1.0	↓	1		155	16 43.1	−77.52
α	TrA	●	1.9	1.4	↓	−4		420	16 48.7	−69.03
η	Pav	•	3.6	1.2	↓	−2		380	17 45.7	−64.72
ζ	Pav	•	4.0	1.1	↓	0		210	18 43.0	−71.43
λ	Pav	♂•	4.2	−.1	│	−4		1500	18 52.2	−62.19
κ	Pav	•	3.9–4.8	0.6	↓	−2		550	18 56.9	−67.23
ε	Pav	•	4.0	0.0	│	1		106	20 00.6	−72.91
δ	Pav	•	3.6	0.8	↓	5		19.9	20 08.7	−66.18
α	Pav	●	1.9	−.1	│	−2 .. Peacock ..		185	20 25.6	−56.74
α	Ind	●	3.1	1.0	↓	1		102	20 37.6	−47.29
β	Pav	●	3.4	0.2	│	0		140	20 45.0	−66.20
β	Ind	●	3.7	1.2	↓	−3		600	20 54.8	−58.45
ϑ	Ind	•	4.4 ☆	0.2	│	2		97	21 19.9	−53.45
ν	Oct	•	3.7	1.0	↓	2		72	21 41.5	−77.39
γ	Gru	●	3.0	−.1	│	−1		200	21 53.9	−37.37
α	Gru	●	1.7	−.1	│	−1 .. Alnair ..		100	22 08.2	−46.96
α	Tuc	●	2.9	1.4	↓	−1		200	22 18.5	−60.26
π	Gru	·	5.2 ☆	0.7	↓	0		133,500	22 23.0	−45.93
δ¹	Gru	•	4.0	1.0	↓	−1 } Sep. 16'.1		340	22 29.3	−43.50
δ²	Gru	•	4.1	1.6	↓	−1		340	22 29.8	−43.75
β	Gru	●	2.1	1.6	↓	−2		170	22 42.7	−46.88
β	Oct	•	4.1	0.2	│	1		142	22 46.1	−81.38
ε	Gru	•	3.5	0.1	│	0		130	22 48.6	−51.32
ι	Gru	•	3.9	1.0	↓	0		190	23 10.4	−45.25
γ	Tuc	•	4.0	0.4	│	2		72	23 17.4	−58.24

BINARY Position V-Mag. B−V Te. Sep. PA Vis.

BINARY		Position	V-Mag.		B−V		Te.	Sep.	PA	Vis.
δ	Aps	•	4.7	5.3	1.7	1.4	↕	103".3		⊡
ϑ	Ind	•	4.5	6.9	0.2	0.6	↕↕	7.0	••	⊡
π	Gru	•	5.6	6.3	0.4	2.2	↓.	258	••	⊡

VARIABLE STAR

VARIABLE	STAR	
κ Pav	⊡ •	⌒
Period	9.095 d	
Max.	2451203	

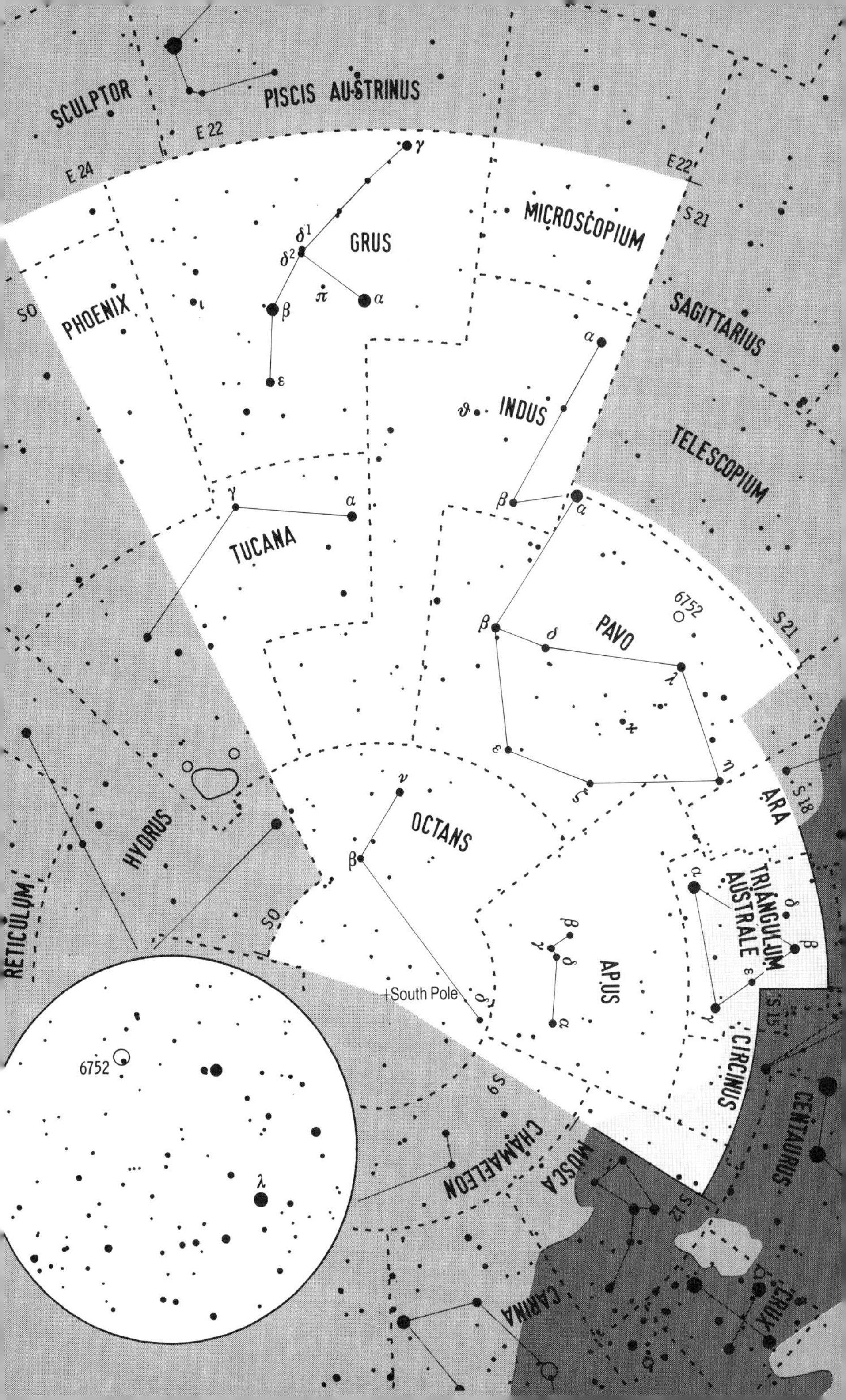
SCULPTOR
PISCIS AUSTRINUS
E 22
E 24
GRUS
MICROSCOPIUM
E22
S21
PHOENIX
S0
SAGITTARIUS
INDUS
TELESCOPIUM
TUCANA
PAVO
6752
S 21
HYDRUS
OCTANS
ARA
S 18
TRIANGULUM AUSTRALE
RETICULUM
S0
APUS
+South Pole
S 15
CIRCINUS
CENTAURUS
6752
S 9
CHAMAELEON
MUSCA
S 12
CRUX
CARINA

Appendix

Brightest Stars ———— Meteor Showers

Brightest Stars		Name	V-Mag.	B–V	Te.	Abs.	Dist.	Chart
		Sun	−26.74	0.66	↓	4.8^{M}	81-min.	
Alpha	Canis Majoris	Sirius	−1.46	0.01	↓	1.4	8.6 ly	E6
Alpha	Carinae ...	Canopus ..	−0.71	0.16	↓	−5.6	310	S3
Alpha	Centauri ...	Toliman[1) ..	−0.28	0.73	✶	4.1	4.40	S15
Alpha	Bootis	Arcturus ..	−0.05	1.23	↓	−0.3	36.5	E15
Alpha	Lyrae	Vega	0.03	0.00	↓	0.6	25.3	N18
Alpha	Aurigae ...	Capella ...	0.07	0.80	↓	−0.5	42	N6
Beta	Orionis	Rigel	0.14	−0.03	✶	−6.9	800	E4
Alpha	Canis Minoris	Procyon ..	0.39	0.42	↓	2.7	11.4	E9
Alpha	Eridani	Achernar ..	0.45	−0.16	↓	−2.8	143	S0
Alpha	Orionis	Betelgeuse .	0.3–0.9	1.84	↓	−4.9	350	E5
Beta	Centauri ...	Hadar, Agena	0.61	−0.24	↓	−5.5	520	S15
Alpha	Crucis	Acrux ...	0.75	−0.24	✶	−4.4	340	S12
Alpha	Aquilae	Altair, Atair	0.77	0.22	↓	2.2	16.7	E21
Alpha	Tauri	Aldebaran .	0.87	1.54	↓	−0.7	66	E3
Alpha	Virginis ...	Spica	0.98	−0.23	↓	−3.5	260	E16
Alpha	Scorpii	Antares ..	0.9–1.1	1.84	✶	−4.8	450	E18
Beta	Geminorum .	Pollux ...	1.14	1.00	↓	1.1	33.5	E7
Alpha	Piscis Austrini	Fomalhaut .	1.16	0.12	↓	1.7	25.2	E22
Beta	Crucis	Mimosa ..	1.25	−0.24	↓	−3.9	340	S12
Alpha	Cygni	Deneb ...	1.25	0.09	↓	−7.7	2 000	N20
Alpha	Leonis ...	Regulus ..	1.36	−0.10	✶	−0.5	77	E11
Epsilon	Canis Majoris	Adhara ...	1.50	−0.21	↓	−4.1	430	E6
Alpha	Geminorum .	Castor ...	1.58	0.03	✶	0.6	52	E7
Gamma	Crucis	Gacrux ...	1.60	1.57	↓	−0.6	88	S12
Lambda	Scorpii	Shaula ...	1.62	−0.23	↓	−4.9	600	S21
Gamma	Orionis	Bellatrix ..	1.64	−0.22	↓	−2.7	240	E5

[1) also called Rigil Kentaurus

Meteor Shower	Start	Max.	End	Time	Rate	Radiant	Speed	Source Comet
Quadrantids	Jan 2	Jan 3	Jan 4	0–7^{h}	40/h	15^{h}20^{m} +50°	45$\frac{km}{s}$	planetar
Lyrids	Apr 15	Apr 22	Apr 24	21–4	10	18 10 +35	50	1861 I
η Aquarids	May 1	May 5	May 10	3–4	15	22 30 0	65	Halley
δ Aquarids	Jul 24	Jul 31	Aug 10	0–3	15	23 00 −15	30	ecliptical
Perseids	Jul 28	Aug 11	Aug 20	21–4	80	3 00 +58	60	Swift-Tuttle
Orionids	Oct 15	Oct 20	Oct 25	23–6	15	6 20 +15	65	Halley
Taurids	Oct 20	Nov 10	Nov 30	19–6	10	3 50 +20	30	Encke
Leonids	Nov 15	Nov 17	Nov 19	1–6	10	10 10 +20	70	Tempel-Tuttle
Geminids	Dec 8	Dec 13	Dec 15	19–7	50	7 30 +30	35	ecliptical
Ursids	Dec 20	Dec 22	Dec 23	18–7	10	14 30 +75	35	Tuttle?

Time: period of visibility for northern-hemisphere observers.
Rate: meteor frequency at maximum, radiant at zenith (zenith hourly rate).

Calendar ———————————— 1999–2008

Year:	99		00		01		02		03		04		05		06		07		08	
Moon	Nw	Fl	Nw	Fl	Nw	Fl	Nw	Fl	Nw	Fl	Nw	Fl	Nw	Fl	Nw	Fl	Nw	Fl	Nw	Fl
January	17	$_{31}^{2}$	6	_21_	24	_9_	13	28	2	18	21	7	10	25	29	14	19	3	8	22
February	16	–	5	19	23	8	12	27	1	16	20	6	8	24	28	13	17	2	7	_21_
March	17	$_{31}^{2}$	6	20	25	9	14	28	3	18	20	6	10	25	_29_	_14_	_19_	_3_	7	21
April	16	30	4	18	23	8	12	27	1	16	_19_	5	_8_	_24_	27	13	17	2	6	20
May	15	30	4	18	23	7	12	_26_	$_{31}^{1}$	_16_	19	_4_	8	23	27	13	16	2	5	20
June	13	28	2	16	_21_	6	_10_	24	29	14	17	3	6	22	25	11	15	$_{30}^{1}$	3	18
July	13	_28_	$_{31}^{1}$	_16_	20	_5_	10	24	29	13	17	$_{31}^{2}$	6	21	25	11	14	30	3	18
August	_11_	26	29	15	19	4	8	22	27	12	16	30	5	19	23	9	12	_28_	$_{30}^{1}$	_16_
September	9	25	27	13	17	2	7	21	26	10	14	28	3	18	_22_	_7_	_11_	26	29	15
October	9	24	27	13	16	2	6	21	25	10	_14_	_28_	_3_	_17_	22	7	11	26	28	14
November	8	23	25	11	15	$_{30}^{1}$	4	_20_	_23_	_9_	12	26	2	16	20	5	9	24	27	13
December	7	22	_25_	11	_14_	_30_	_4_	19	23	8	12	26	$_{31}^{1}$	15	20	5	9	24	27	12

	99	00	01	02	03	04	05	06	07	08
Mercury e.	Mar 3	Feb 15	Jan 28	Jan 11	–	–	Mar 12	Feb 24	Feb 7	Jan 22
‹ west	Apr 16	Mar 28	Mar 11	Feb 21	Feb 4	Jan 17	Apr 26	Apr 8	Mar 22	Mar 3
› east	Jun 28	Jun 9	May 22	May 4	Apr 16	Mar 29	Jul 9	Jun 20	Jun 2	May 14
‹ west	Aug 14	Jul 27	Jul 9	Jun 21	Jun 3	May 14	Aug 23	Aug 7	Jul 20	Jul 1
› east	Oct 24	Oct 6	Sep 18	Sep 1	Aug 14	Jul 27	Nov 3	Oct 17	Sep 29	Sep 11
‹ west	Dec 3	Nov 15	Oct 29	Oct 13	Sep 27	Sep 9	Dec 12	Nov 25	Nov 8	Oct 22
› east	–	–	–	Dec 26	Dec 9	Nov 21	–	–	–	–
‹ west	–	–	–	–	–	Dec 29	–	–	–	–
Venus ⅾ east	Jun 11	–	Jan 17	Aug 22	–	Mar 29	Nov 3	–	Jun 9	–
inf.	Aug 20	–	Mar 30	Oct 31	–	Jun 8	–	Jan 13	Aug 18	–
ⅾ west	Oct 30	–	Jun 8	–	Jan 11	Aug 17	–	Mar 25	Oct 28	–
° sup.	–	Jun 11	–	Jan 14	Aug 18	–	Mar 31	Oct 27	–	Jun 9
Mars op.	Apr 24	–	Jun 13	–	Aug 28	–	Nov 7	–	Dec 24	–
constellation	Vir ○	–	Oph○	–	Aqr ○	–	Ari ○	–	Gem○	–
Jupiter op.	Oct 23	Nov 28	–	Jan 1	Feb 2	Mar 4	Apr 3	May 4	Jun 6	Jul 9
constellation	Psc○	Tau○	–	Gem○	Cnc○	Leo ○	Vir ○	Lib ○	Oph○	Sgr ○
Saturn op.	Nov 6	Nov 19	Dec 3	Dec 17	Dec 31	–	Jan 13	Jan 27	Feb 10	Feb 24
constellation	Ari ⊶	Tau⊙	Tau⊙	Tau⊙	Gem⊙	–	Gem⊙	Cnc⊙	Leo ⊶	Leo ⊶

First Sunday J D	2451	2451	2451/2	2452	2452	2453	2453	2453/4	2454	2454
January	3 179	2 544	7 910	6 275	5 640	4 005	2 371	1 736	7 101	6 466
February	7 210	6 575	4 941	3 306	2 671	1 036	6 402	5 767	4 132	3 497
March	7 238	5 604	4 969	3 334	2 699	7 065	6 430	5 795	4 160	2 526
April	4 269	2 635	1 000	7 365	6 730	4 096	3 461	2 826	1 191	6 557
May	2 299	7 665	6 030	5 395	4 760	2 126	1 491	7 856	6 221	4 587
June	6 330	4 696	3 061	2 426	1 791	6 157	5 522	4 887	3 252	1 618
July	4 360	2 726	1 091	7 456	6 821	4 187	3 552	2 917	1 282	6 648
August	1 391	6 757	5 122	4 487	3 852	1 218	7 583	6 948	5 313	3 679
September	5 422	3 788	2 153	1 518	7 883	5 249	4 614	3 979	2 344	7 710
October	3 452	1 818	7 183	6 548	5 913	3 279	2 644	1 009	7 374	5 740
November	7 483	5 849	4 214	3 579	2 944	7 310	6 675	5 040	4 405	2 771
December	5 513	3 879	2 244	1 609	7 974	5 340	4 705	3 070	2 435	7 801

New Moon, Full Moon: underscored are <u>solar and lunar eclipses</u>, <u>total eclipses</u>.

Mercury, Venus: greatest elongation east/west, inferior/superior conjunction.

Mars, Jupiter, Saturn: date of opposition, position, and planetary disk.

2009–2018 ——————————————— Calendar

Year:	09	10	11	12	13	14	15	16	17	18
Moon	Nw Fl	Nw Fl	Nw Fl	Nw Fl	Nw Fl	Nw Fl	Nw Fl	Nw Fl	Nw Fl	Nw Fl
January	26 11	15 30	4 19	23 9	11 27	30^{1} 16	20 5	10 24	28 12	17 31^{2}
February	25 9	14 28	3 18	21 7	10 25	- 14	18 3	8 22	26 11	15 -
March	26 11	15 30	4 19	22 8	11 27	30^{1} 16	20 5	9 23	28 12	17 31^{2}
April	25 9	14 28	3 18	21 6	10 25	29 15	18 4	7 22	26 11	16 30
May	24 9	14 27	3 17	20 6	10 25	28 14	18 4	6 21	25 10	15 29
June	22 7	12 26	1 15	19 4	8 23	27 13	16 2	5 20	24 9	13 28
July	22 7	11 26	30^{1} 15	19 3	8 22	26 12	16 31^{2}	4 19	23 9	13 27
August	20 6	10 24	29 13	17 31^{2}	6 21	25 10	14 29	2 18	21 7	11 26
September	18 4	8 23	27 12	16 30	5 19	24 9	13 28	1 16	20 6	9 25
October	18 4	7 23	26 12	15 29	5 18	23 8	13 27	30^{1} 16	19 5	9 24
November	16 2	6 21	25 10	13 28	3 17	22 6	11 25	29 14	18 4	7 23
December	16 31^{2}	5 21	24 10	13 28	3 17	22 6	11 25	29 14	18 3	7 22

Mercury e.	Jan 4	-	-	Mar 5	Feb 16	Jan 31	Jan 14	-	-	-
‹ west	Feb 13	Jan 27	Jan 9	Apr 18	Mar 31	Mar 14	Feb 24	Feb 7	Jan 19	Jan 1
› east	Apr 26	Apr 8	Mar 23	Jul 1	Jun 12	May 25	May 7	Apr 18	Apr 1	Mar 15
‹ west	Jun 13	May 26	May 7	Aug 16	Jul 30	Jul 12	Jun 24	Jun 5	May 17	Apr 29
› east	Aug 24	Aug 7	Jul 20	Oct 26	Oct 9	Sep 21	Sep 4	Aug 16	Jul 30	Jul 12
‹ west	Oct 6	Sep 19	Sep 3	Dec 4	Nov 18	Nov 1	Oct 16	Sep 28	Sep 12	Aug 26
› east	Dec 18	Dec 1	Nov 14	-	-	-	Dec 29	Dec 11	Nov 24	Nov 6
‹ west	-	-	Dec 23	-	-	-	-	-	-	Dec 15
Venus☽ east	Jan 14	Aug 20	-	Mar 27	Nov 1	-	Jun 6	-	Jan 12	Aug 17
inf.	Mar 27	Oct 29	-	Jun 6	-	Jan 11	Aug 15	-	Mar 25	Oct 26
☾ west	Jun 5	-	Jan 8	Aug 15	-	Mar 22	Oct 26	-	Jun 3	-
° sup.	-	Jan 11	Aug 16	-	Mar 28	Oct 25	-	Jun 6	-	Jan 9
Mars op.	-	Jan 29	-	Mar 3	-	Apr 8	-	May 22	-	Jul 27
constellation	-	Cnc ○	-	Leo ○	-	Vir ○	-	Sco ○	-	Cap ○
Jupiter op.	Aug 14	Sep 21	Oct 29	Dec 3	-	Jan 5	Feb 6	Mar 8	Apr 7	May 9
constellation	Cap○	Psc ○	Ari ○	Tau○	-	Gem○	Cnc○	Leo ○	Vir ○	Lib ○
Saturn op.	Mar 8	Mar 22	Apr 3	Apr 15	Apr 28	May 10	May 23	Jun 3	Jun 15	Jun 27
constellation	Leo ⊕	Vir ⊕	Vir ⊕	Vir ⊕	Lib ⊕	Lib ⊕	Lib ⊕	Oph⊕	Oph⊕	Sgr ⊕

First Sunday	J D	2454/5	2455	2455	2455/6	2456	2456	2457	2457	2457/8	2458
January		4 832	3 197	2 562	1 927	6 293	5 658	4 023	3 388	1 754	7 119
February		1 863	7 228	6 593	5 958	3 324	2 689	1 054	7 419	5 785	4 150
March		1 891	7 256	6 621	4 987	3 352	2 717	1 082	6 448	5 813	4 178
April		5 922	4 287	3 652	1 018	7 383	6 748	5 113	3 479	2 844	1 209
May		3 952	2 317	1 682	6 048	5 413	4 778	3 143	1 509	7 874	6 239
June		7 983	6 348	5 713	3 079	2 444	1 809	7 174	5 540	4 905	3 270
July		5 013	4 378	3 743	1 109	7 474	6 839	5 204	3 570	2 935	1 300
August		2 044	1 409	7 774	5 140	4 505	3 870	2 235	7 601	6 966	5 331
September		6 075	5 440	4 805	2 171	1 536	7 901	6 266	4 632	3 997	2 362
October		4 105	3 470	2 835	7 201	6 566	5 931	4 296	2 662	1 027	7 392
November		1 136	7 501	6 866	4 232	3 597	2 962	1 327	6 693	5 058	4 423
December		6 166	5 531	4 896	2 262	1 627	7 992	6 357	4 723	3 088	2 453

First Sunday: left entry in table. **JD, Julian Date:** right entry in table.

JD = thousand (head of column) + entry + day of month + $(UT - 12^{h}) / 24^{h}$

Example: 25 Dec. 2010, 1^{h}: JD = 2455000 + 531 + 25 − $\frac{11}{24}$ = 2455555.542.

Nebula Numbers

M = Messier, Chart NGC = New General Catalogue, Chart Number

M	Chart	M	Chart	NGC	Chart	NGC	Chart	NGC	Chart	NGC	Chart	NGC	Chart
M 1	E3	M 56	N18	55	S0	2068	E5	3242	E10	4736	N12	6543	N16
M 2	E24	M 57	N18	104	S0	2070	S3	3293	S9	4755	S12	6572	E19
M 3	E15	M 58	E14	205	N0	2099	N6	3351	E11	4762	E14	6611	E20
M 4	E18	M 59	E14	221	N0	2129	E7	3368	E11	4826	E13	6613	E20
M 5	E15	M 60	E14	224	N0	2168	E7	3372	S9	5024	E13	6618	E20
M 6	E18	M 61	E14	247	E0	2175	E7	3379	E11	5055	N12	6626	E20
M 7	E18	M 62	E18	253	E0	2237	E9	3384	E11	5128	S15	6633	E19
M 8	E20	M 63	N12	281	N2	2244	E9	3532	S9	5139	S15	6637	E20
M 9	E17	M 64	E13	288	E0	2261	E7	3556	N10	5194	N12	6656	E20
M 10	E17	M 65	E11	292	S0	2264	E7	3587	N10	5195	N12	6681	E20
M 11	E19	M 66	E11	362	S0	2281	N6	3623	E11	5236	E16	6694	E19
M 12	E17	M 67	E9	457	N2	2287	E6	3627	E11	5272	E15	6705	E19
M 13	N14	M 68	E12	559	N2	2301	E9	3628	E11	5457	N10	6712	E19
M 14	E17	M 69	E20	581	N2	2323	E8	3766	S12	5460	S15	6715	E20
M 15	E23	M 70	E20	598	N0	2324	E9	3992	N10	5746	E15	6720	N18
M 16	E20	M 71	E21	628	E1	2359	E8	4192	E14	5822	S15	6723	S21
M 17	E20	M 72	E24	650	N0	2360	E8	4216	E14	5866	N16	6752	S24
M 18	E20	M 73	E24	654	N2	2362	E6	4244	N12	5904	E15	6779	N18
M 19	E18	M 74	E1	663	N2	2392	E7	4254	E14	5907	N16	6809	E22
M 20	E20	M 75	E22	752	N0	2403	N8	4258	N12	5986	S18	6818	E22
M 21	E20	M 76	N0	869	N2	2422	E8	4303	E14	6067	S18	6822	E22
M 22	E20	M 77	E0	884	N2	2423	E8	4321	E14	6087	S18	6826	N18
M 23	E20	M 78	E5	891	N0	2437	E8	4361	E12	6093	E18	6838	E21
M 24	E20	M 79	E4	1023	N4	2438	E8	4374	E14	6121	E18	6853	E21
M 25	E20	M 80	E18	1039	N4	2447	E6	4382	E14	6124	S21	6864	E22
M 26	E19	M 81	N8	1068	E0	2451	S6	4406	E14	6171	E17	6913	N20
M 27	E21	M 82	N8	1245	N4	2477	S6	4449	N12	6205	N14	6939	N22
M 28	E20	M 83	E16	1291	S0	2516	S3	4472	E14	6210	E19	6940	N20
M 29	N20	M 84	E14	1316	S0	2539	E8	4486	E14	6218	E17	6946	N20
M 30	E22	M 85	E14	1365	S0	2546	S6	4490	N12	6231	S21	6960	N20
M 31	N0	M 86	E14	1491	N4	2547	S6	4494	E13	6254	E17	6981	E24
M 32	N0	M 87	E14	1528	N4	2548	E10	4501	E14	6266	E18	6992	N20
M 33	N0	M 88	E14	1535	E2	2632	E9	4526	E14	6273	E18	6994	E24
M 34	N4	M 89	E14	1647	E3	2682	E9	4548	E14	6333	E17	7000	N20
M 35	E7	M 90	E14	1788	E5	2683	N8	4552	E14	6341	N14	7009	E24
M 36	N6	M 91	E14	1851	S3	2808	S9	4559	E13	6369	E18	7027	N20
M 37	N6	M 92	N14	1904	E4	2841	N8	4565	E13	6388	S21	7078	E23
M 38	N6	M 93	E6	1912	N6	2903	E11	4569	E14	6397	S18	7089	E24
M 39	N24	M 94	N12	1931	N6	2976	N8	4579	E14	6402	E17	7092	N24
M 40	N10	M 95	E11	1952	E3	3031	N8	4590	E12	6405	E18	7099	E22
M 41	E6	M 96	E11	1960	N6	3034	N8	4594	E12	6475	E18	7209	N24
M 42	E4	M 97	N10	1973	E4	3077	N8	4621	E14	6494	E20	7243	N24
M 43	E4	M 98	E14	1976	E4	3114	S9	4631	E13	6503	N16	7293	E24
M 44	E9	M 99	E14	1981	E4	3115	E10	4649	E14	6514	E20	7331	E23
M 45	E3	M 100	E14	1982	E4	3132	S6	4656	E13	6523	E20	7654	N22
M 46	E8	M 101	N10	2024	E5	3184	N10	4697	E12	6531	E20	7662	N24
M 47	E8	M 102	N16			3201	S6	4725	E13	6541	S21	7789	N22
M 48	E10	M 103	N2										
M 49	E14	M 104	E12	IC 1396	N22								
M 50	E8	M 105	E11	IC 2391	S6								
M 51	N12	M 106	N12	IC 2602	S9								
M 52	N22	M 107	E17	IC 4665	E17								
M 53	E13	M 108	N10	IC 4725	E20								
M 54	E20	M 109	N10	IC 4756	E19								
M 55	E22	M 110	N0	IC 5067	N20								

η Carinae Nebula	= NGC 3372	S9
o Velorum Cluster =	IC 2391	S6
ω Centauri	= NGC 5139	S15
47 Tucanae	= NGC 104	S0
Centaurus A	= NGC 5128	S15
Fornax A	= NGC 1316	S0
Virgo A	= M 87	E14

Nebula Names

Nebula Name	NGC	Messier	Const.	Mag.	Type	Vis.	Chart
Andromeda Galaxy	224	M 31	And	4	Glx	🎲	N0
Barnard's Galaxy	6822	...	Sgr	9	Glx	🎲	E22
Black Eye Galaxy	4826	M 64	Com	9	Glx	🎲	E13
Blinking Planetary	6826	...	Cyg	8½	PN	🎲	N18
Blue Snowball	7662	...	And	8½	PN	🎲	N24
Butterfly Cluster	6405	M 6	Sco	4½	OC	🎲	E18
Christmas Tree (Cluster)	2264	...	Mon	4	OC	🎲	E7
Coalsack			Cru	(3)	Dark N.	🎲	S12
Coma (Star) Cluster			Com	2½	OC	🎲	E13
Crab Nebula	1952	M 1	Tau	8	DN	🎲	E3
Double Cluster, h and χ Persei	869, 884		Per	4	OC	🎲	N2
Dumbbell Nebula	6853	M 27	Vul	7	PN	🎲	E21
Eagle Nebula	6611	M 16	Ser	6	DN	🎲	E20
Eskimo Nebula	2392	...	Gem	9	PN	🎲	E7
Fornax (Galaxy) Cluster	1316, 1365		For	9	Glx	🎲	S0
Ghost of Jupiter	3242	...	Hya	8	PN	🎲	E10
Helix Nebula	7293	...	Aqr	7	PN	🎲	E24
Hercules Cluster	6205	M 13	Her	6	GC	🎲	N14
Hubble's Variable Nebula	2261	...	Mon	9½	DN	🎲	E7
Hyades			Tau	1	OC	🎲	E3
Jewel Box, κ Crucis (Cluster)	4755	...	Cru	4½	OC	🎲	S12
Lagoon Nebula	6523	M 8	Sgr	4½	DN	🎲	E20
Large Magellanic Cloud, LMC			Dor	0	Glx	🎲	S3
Little Dumbbell	650	M 76	Per	10	PN	🎲	N0
Makarian's (Galaxy) Chain	M 86–M 88		Com	9½	Glx	🎲	E14
North America Nebula	7000	...	Cyg	5	DN	🎲	N20
Omega Nebula, Swan Nebula	6618	M 17	Sgr	6	DN	🎲	E20
Orion Nebula	1976	M 42	Ori	3½	DN	🎲	E4
Owl Nebula	3587	M 97	UMa	10	PN	🎲	N10
Pelican Nebula	IC 5067	...	Cyg	7	DN	🎲	N20
Pinwheel Galaxy	5457	M 101	UMa	8	Glx	🎲	N10
Pleiades, Seven Sisters		M 45	Tau	1½	OC	🎲	E3
Praesepe, Beehive (Cluster)	2632	M 44	Cnc	3½	OC	🎲	E9
Quasi-Stellar Object 3 C 273			Vir	13	Quasar	🎲	E14
Ring Nebula	6720	M 57	Lyr	8½	PN	🎲	N18
Rosette Nebula	2237	...	Mon	6	DN	🎲	E9
Saturn Nebula	7009	...	Aqr	8	PN	🎲	E24
Sculptor Galaxy	253	...	Scl	7½	Glx	🎲	E0
Small Magellanic Cloud, SMC			Tuc	2½	Glx	🎲	S0
Sombrero Galaxy	4594	M 104	Vir	8½	Glx	🎲	E12
Southern Pleiades	IC 2602	...	Car	2	OC	🎲	S9
Spindle Galaxy	3115	...	Sex	9½	Glx	🎲	E10
Tarantula Nebula	2070	...	Dor	4½	DN	🎲	S3
Triangulum Galaxy	598	M 33	Tri	6	Glx	🎲	N0
Trifid Nebula	6514	M 20	Sgr	7	DN	🎲	E20
Veil Nebula, ⎰Filametary Nebula	6960	...	Cyg	9	DN	🎲	N20
Cirrus N., ⎱Network Nebula	6992	...	Cyg	7½	DN	🎲	N20
Virgo Cluster			Vir	8½	Glx	🎲	E14
Whirlpool Galaxy	5194	M 51	CVn	8½	Glx	🎲	N12

Star Names

Star Name	Designat.	Mag.	Chart
Acamar ...	ϑ Eri •	2.9 ✭	S0
Achernar ...	α Eri ●	0.5	S0
Acrab	β Sco •	2.4 ✭	E18
Acrux	α Cru ●	0.7 ✭	S12
Acubens ...	α Cnc ·	4.3	E9
Adhara	ε CMa●	1.5 ✭	E6
Agena	β Cen ●	0.6	S15
Alamak	γ And •	2.1 ✭	N0
Albireo	β Cyg •	2.9 ✭	N18
Alchiba	α Crv ·	4.0	E12
Alcor	80 UMa•	4.0	N10
Alcyone ...	η Tau •	2.8	E3
Aldebaran ..	α Tau ●	0.9	E3
Alderamin ..	α Cep •	2.5	N22
Aldhafera ..	ζ Leo •	3.4	E11
Alfirk	β Cep •	3.2 ✭	N22
Algenib ...	γ Peg •	2.8	E1
Algieba	γ Leo ●	2.0 ✭	E11
Algiedi	α Cap •	3.1 ✭	E22
Algol	β Per •	2.1–3.4	N4
Algorab ...	δ Crv •	2.9	E12
Alhena	γ Gem●	1.9	E7
Alioth	ε UMa●	1.8	N10
Alkaid	η UMa●	1.9	N10
Alkalurops ..	μ Boo ·	4.2 ✭	N14
Alkes	α Crt ·	4.1	E12
Alnair	α Gru ●	1.7	S24
Alnasl	γ Sgr •	3.0	E20
Alnilam ...	ε Ori ●	1.7	E5
Alnitak	ζ Ori ●	1.7 ✭	E5
Alphard ...	α Hya ●	2.0	E10
Alphekka ...	α CrB ●	2.2	E15
Alpheratz ..	α And ●	2.1	N0
Alschain ...	β Aql •	3.7	E21
Altair	α Aql ●	0.8	E21
Altais	δ Dra •	3.1	N16
Altarf	β Cnc •	3.5	E9
Alterf	λ Leo ·	4.3	E11
Aludra	η CMa●	2.4 ✭	E6
Alula Australis	ξ UMa•	3.8 ✭	N12
Alula Borealis	ν UMa•	3.5	N12
Alya	ϑ Ser ·	4.0 ✭	E19
Ankaa	α Phe •	2.4	S0
Antares ...	α Sco ●	0.9–1.1 ✭	E18
Arcturus ...	α Boo ●	0.0	E15
Arneb	α Lep •	2.6	E4
Asellus Australis	δ Cnc ·	3.9	E9

Star Name	Designat.	Mag.	Chart
Asellus Borealis	γ Cnc ·	4.7	E9
Aspidiske .	ξ Pup •	3.2 ✭	E6
Atair	α Aql ●	0.8	E21
Atik	o Per •	3.8	N4
Atlas	27 Tau •	3.6	E3
Avoir	ε Car •	1.9	S3
Baham ...	ϑ Peg •	3.5	E23
Barnard's Star	†	9.5	E17
Baten Kaitos	ζ Cet •	3.7	E0
Bellatrix ..	γ Ori ●	1.6	E5
Benetnasch	η UMa●	1.9	N10
Betelgeuse .	α Ori ●	0.3–0.9	E5
Canopus ..	α Car ●	−0.7	S3
Capella ...	α Aur ●	0.1	N6
Castor ...	α Gem●	1.6 ✭	E7
Cebalrai ..	β Oph •	2.8	E17
Ceginus ..	γ Boo •	3.0	N14
Chaph ...	β Cas •	2.3	N2
Cor Caroli .	α CVn •	2.8 ✭	N12
Coxa	ϑ Leo •	3.3	E11
Cursa	β Eri •	2.8	E2
Deneb ...	α Cyg ●	1.3	N20
Deneb Algedi	δ Cap •	2.8–3.1	E22
Deneb Kaitos	β Cet ●	2.0	E0
Denebola ..	β Leo ●	2.1	E11
Diadem ...	α Com ·	4.3	E13
Diphda ...	β Cet ●	2.0	E0
Double Double	ε Lyr ·	3.9 ✭	N18
Dubhe ...	α UMa●	1.8	N10
Edasich ..	ι Dra •	3.3	N16
Electra ...	17 Tau •	3.7	E3
Elmuthalleth	α Tri •	3.4	N0
Elnath ...	β Tau ●	1.7	E3
Enif	ε Peg •	2.4	E23
Errai ...	γ Cep •	3.2	N22
Ettanin ...	γ Dra •	2.2	N16
Fomalhaut .	α PsA ●	1.2	E22
Gacrux ...	γ Cru ●	1.6 ✭	S12
Gemma ..	α CrB ●	2.2	E15
Giauzar ..	λ Dra •	3.8	N16
Gienah ...	γ Crv •	2.6	E12
Gomeisa ..	β CMi •	2.9	E9
Grumium .	ξ Dra ·	3.7	N16
Hadar ...	β Cen ●	0.6	S15
Hamal ...	α Ari ●	2.0	E1
Homam ..	ζ Peg •	3.4	E23
Izar	ε Boo •	2.4 ✭	E15

Star Names

Star Name	Designat.	Mag.	Chart
Kaus Australis	ε Sgr ●	1.8	E20
Kaus Borealis	λ Sgr ●	2.8	E20
Kaus Media	δ Sgr ●	2.7	E20
Kitalphar	α Equ •	3.9	E23
Kochab . . .	β UMi ●	2.1	NP
La Superba	Y CVn ·	5.2–5.6	N12
Lesath . . .	υ Sco ●	2.7	S21
Maia	20 Tau •	3.8	E3
Marfik . . .	λ Oph •	3.8 ✲	E17
Markab . . .	α Peg ●	2.5	E23
Matar . . .	η Peg ●	2.9	E23
Mebsuta . .	ε Gem ●	3.1	E7
Megrez . . .	δ UMa●	3.3	N10
Mekbuda . .	ζ Gem •3.6–4.2✲		E7
Menkalinan	β Aur ●	1.9	N6
Menkar . . .	α Cet ●	2.5	E0
Menkib . . .	ξ Per •	4.0	N4
Merak . . .	β UMa●	2.3	N10
Merope . . .	23 Tau •	4.1	E3
Mesarthim .	γ Ari ●	3.9 ✲	E1
Miaplacidus	β Car ●	1.7	S9
Mimosa . .	β Cru ●	1.3	S12
Mintaka . .	δ Ori ●	2.2 ✲	E5
Mira	o Cet ·	3.4–9.2	E0
Mirach . . .	β And ●	2.1	N0
Mirphak . .	α Per ●	1.8	N4
Mirzam . . .	β CMa●	2.0	E6
Mizar	ζ UMa●	2.0 ✲	N10
Muphrid . .	η Boo ●	2.7	E15
Nath	β Tau ●	1.7	E3
Nekkar . . .	β Boo •	3.5	N14
Nihal . . .	β Lep ●	2.8	E4
North Star .	α UMi●	2.0	NP
Nunki . . .	σ Sgr ●	2.0	E20
Nusakan . .	β CrB •	3.7	E15
Peacock . .	α Pav ●	1.9	S24
Phact	α Col ●	2.6	E2
Phad	γ UMa●	2.4	N10
Phegda . . .	γ UMa●	2.4	N10
Pherkad . .	γ UMi •	3.0	NP
Phurud . . .	ζ CMa●	3.0 ✲	E6
Piazzi's Flying	61 Cyg ·	4.8 ✲	N24
Pleione ⌊Star⌋	28 Tau ·	4.9–5.2	E3
Polaris . . .	α UMi●	2.0	NP
Pollux . . .	β Gem●	1.1	E7
Porrima . .	γ Vir •	2.7 ✲	E12
Procyon . .	α CMi●	0.4	E9
Pulcherrima	ε Boo ●	2.4 ✲	E15
Rasalgethi .	α Her •2.6–3.4✲		E19

Star Name	Designat.	Mag.	Chart
Rasalhague .	α Oph ●	2.1	E17
Rastaben . . .	β Dra ●	2.8	N16
Regulus . . .	α Leo ●	1.4 ✲	E11
Rigel	β Ori ●	0.1 ✲	E4
Rigil Kentaurus	α Cen ●−0.3 ✲		S15
Ruchbah . . .	δ Cas ●	2.7	N2
Ruticulus . .	β Her ●	2.8	E19
Sabik	η Oph ●	2.4	E17
Sadachbia . .	γ Aqr •	3.9	E24
Sadalbari . . .	μ Peg •	3.5	E23
Sadalmelik . .	α Aqr ●	3.0	E24
Sadalsuud . .	β Aqr ●	2.9	E24
Sadr	γ Cyg ●	2.2	N20
Saiph	κ Ori ●	2.1	E4
Sargas	ϑ Sco ●	1.9	S21
Scheat	β Peg ● 2.4–2.6		E23
Schedir	α Cas ●	2.2	N2
Shaula	λ Sco ●	1.6	S21
Sheliak	β Lyr •3.3–4.2✲		N18
Sheratan . . .	β Ari ●	2.6	E1
Sirius	α CMa●−1.5		E6
Sirrah	α And ●	2.1	N0
Spica	α Vir ●	1.0	E16
Suhail Al Muhlif	γ Vel ●1.5–1.7✲		S6
Suhail Al Wazn	λ Vel ●	2.2	S6
Sulaphat . . .	γ Lyr •	3.2	N18
Talitha	ι UMa●	3.1	N8
Tania Australis	μ UMa●	3.1	N10
Tania Borealis	λ UMa●	3.5	N10
Tarazed . . .	γ Aql ●	2.7	E21
Taygeta . . .	19 Tau ·	4.3	E3
Tejat Posterior	μ Gem ●	2.9	E7
Tejat Prior . .	η Gem •3.2–3.4		E7
Thuban . . .	α Dra •	3.7	N16
Toliman . . .	α Cen ●−0.3 ✲		S15
Trapezium . .	ϑ¹ Ori ·	4.6 ✲	E4
Unukalhai . .	α Ser ●	2.6	E16
Vega	α Lyr ●	0.0	N18
Vindemiatrix	ε Vir ●	2.8	E16
Wasat	δ Gem●	3.5	E7
Wezen	δ CMa●	1.8	E6
Yed Posterior	ε Oph ●	3.2	E17
Yed Prior . .	δ Oph ●	2.7	E17
Zaniah	η Vir •	3.9	E12
Zaurak . . .	γ Eri •	3.0	E2
Zawijava . . .	β Vir •	3.6	E12
Zosma	δ Leo ●	2.6	E11
Zubenelgenubi	α Lib ●	2.6 ✲	E16
Zubeneschemali	β Lib ●	2.6	E16

Constellations

Abbr.	Constellation	Genitive	Meaning	Chart	Neb.	St.
And	Andromeda	Andromedae	Chained Lady	N0 (N24)	6	16
Ant	Antlia	Antliae	Air Pump	E10	-	1
Aps	Apus	Apodis	Bird of Paradise	S24	-	4
Aqr	Aquarius	Aquarii	Water-bearer	E24	5	19
Aql	Aquila	Aquilae	Eagle	E21	-	15
Ara	Ara	Arae	Altar	S18	1	8
Ari	Aries	Arietis	Ram	E1	-	8
Aur	Auriga	Aurigae	Charioteer	N6	5	14
Boo	Bootes	Bootis	Herdsman	E15 (N14)	-	18
Cae	Caelum	Caeli	Chisel	S3	-	-
Cam	Camelopardalis	Camelopadalis	Giraffe	N2 (NP)	1	6
Cnc	Cancer	Cancri	Crab	E9	2	11
CVn	Canes Venatici	Canum Venaticorum	Hunting Dogs	N12	11	5
CMa	Canis Major	Canis Majoris	Big Dog	E6	4	18
CMi	Canis Minor	Canis Minoris	Little Dog	E9	-	3
Cap	Capricornus	Capricorni	Sea Goat	E22	1	12
Car	Carina	Carinae	Keel	S9 (S3)	7	20
Cas	Cassiopeia	Cassiopeiae	Queen in the Chair	N2	8	17
Cen	Centaurus	Centauri	Centaur	S15 (S12)	4	26
Cep	Cepheus	Cephei	Monarch	N22	3	13
Cet	Cetus	Ceti	Whale	E0	2	18
Cha	Chamaeleon	Chamaeleontis	Chameleon	S9	-	4
Cir	Circinus	Circini	Compasses	S15	-	4
Col	Columba	Columbae	Dove	E2 (S3)	1	6
Com	Coma Berenices	Comae Berenices	Berenice's Hair	E13	13	8
CrA	Corona Australis	Coronae Australis	Southern Crown	S21	1	7
CrB	Corona Borealis	Coronae Borealis	Northern Crown	E15	-	10
Crv	Corvus	Corvi	Crow	E12	1	5
Crt	Crater	Crateris	Cup	E12	-	4
Cru	Crux	Crucis	Southern Cross	S12	2	7
Cyg	Cygnus	Cygni	Swan	N20 (N18)	8	31
Del	Delphinus	Delphini	Dolphin	E21	-	5
Dor	Dorado	Doradus	Gold Fish	S3	2	3
Dra	Draco	Draconis	Dragon	N16	4	21
Equ	Equuleus	Equulei	Little Horse	E23	-	4
Eri	Eridanus	Eridani	River	E2 (S0)	2	27
For	Fornax	Fornacis	Furnace	E2	3	3
Gem	Gemini	Geminorum	Twins	E7	3	17
Gru	Grus	Gruis	Crane (bird)	S24	-	8
Her	Hercules	Herculis	Kneeler	N14 (E19)	3	23
Hor	Horologium	Horologii	Clock	S3 (S0)	-	2
Hya	Hydra	Hydrae	Water Snake	E10 (E12)	4	24
Hyi	Hydrus	Hydri	Little Water Snake	S0	-	3
Ind	Indus	Indi	Indian	S24	-	3
Lac	Lacerta	Lacertae	Lizard	N24	2	7
Leo	Leo	Leonis	Lion	E11	8	17
LMi	Leo Minor	Leonis Minoris	Little Lion	N8	-	2

Abbr.	Constellation	Genitive	Meaning	Chart	Neb.	St.
Lep	Lepus	Leporis	Hare	E4	1	11
Lib	Libra	Librae	Balance (Scales)	E16	-	8
Lup	Lupus	Lupi	Wolf	S18	2	15
Lyn	Lynx	Lyncis	Lynx	N8 (N6)	1	8
Lyr	Lyra	Lyrae	Lyre	N18	2	8
Men	Mensa	Mensae	Table Mountain	S9	-	-
Mic	Microscopium	Microscopii	Microscope	E22	-	-
Mon	Monoceros	Monocerotis	Unicorn	E8 (E9)	7	9
Mus	Musca	Muscae	Fly	S12	-	6
Nor	Norma	Normae	Square, Rule	S18	2	2
Oct	Octans	Octantis	Octant	S24	-	3
Oph	Ophiuchus	Ophiuchi	Serpent-bearer	E17	11	21
Ori	Orion	Orionis	Hunter	E5 (E4)	8	24
Pav	Pavo	Pavonis	Peacock	S24	1	8
Peg	Pegasus	Pegasi	Winged Horse	E23	2	16
Per	Perseus	Persei	Rescuer	N4	8	19
Phe	Phoenix	Phoenicis	Phoenix	S0	-	7
Pic	Pictor	Pictoris	Painter	S3	-	4
Psc	Pisces	Piscium	Fish (two)	E1 (E23)	1	19
PsA	Piscis Austrinus	Piscis Austrini	Southern Fish	E22	-	6
Pup	Puppis	Puppis	Stern	S6 (E6)	9	18
Pyx	Pyxis	Pyxidis	Compass	E10	-	3
Ret	Reticulum	Reticuli	Reticle, Net	S3	-	4
Sge	Sagitta	Sagittae	Arrow	E21	1	5
Sgr	Sagittarius	Sagittarii	Archer	E20	18	16
Sco	Scorpius	Scorpii	Scorpion	E18 (S21)	7	22
Scl	Sculptor	Sculptoris	Sculptor	E0	3	3
Sct	Scutum	Scuti	Shield	E19	3	3
Ser	Serpens (Caput)	Serpentis (Caputis)	Serpent (Head)	E16	1	8
	(Cauda)	(Caudae)	(Tail)	E19	2	4
Sex	Sextans	Sextantis	Sextant	E10	1	3
Tau	Taurus	Tauri	Bull	E3	4	24
Tel	Telescopium	Telescopii	Telescope	S21	-	2
Tri	Triangulum	Trianguli	Triangle	N0	1	6
TrA	Triangulum Aus-	Trianguli Australis	Southern Triangle	S24	-	5
Tuc	Tucana ⌊trale⌋	Tucanae	Toucan	S0 (S24)	3	3
UMa	Ursa Major	Ursae Majoris	Great Bear	N10 (N8)	11	24
UMi	Ursa Minor	Ursae Minoris	Little Bear	NP	-	8
Vel	Vela	Velorum	Sails	S6	4	16
Vir	Virgo	Virginis	Maiden	E16	17	13
Vol	Volans	Volantis	Flying Fish	S3	-	6
Vul	Vulpecula	Vulpeculae	Fox	E21	2	3
88	88 (89)	88 (89)	88 (89)	48	250	900

Chart: chart number(s) for the main part of the constellation.

Neb., St.:   number of nebulae and stars in the catalog.

Glossary

Page numbers locate more detailed explanations.

Abs., Absolute Magnitude (p. 14): The V-magnitude of a star in a distance of 32.6 light years (= 10 parsec). The listed values refer to the maximum brightness for variable stars and to the combined brightness for binaries. The absolute magnitude of the Sun is $4^M.8$.

✷ **Binary** (p. 15): A stellar object to the unaided eye but resolved into two (or more) stars in a telescope. The components are within five arcminutes of each other. If both components appear close to each other but are physically well separated, both distances are listed in the catalog of stars, and the separation is rounded to full arc-seconds. The catalog contains 250 binaries with components brighter than magnitude 8.0.

B–V Color Index (p. 13 and Fig. p. 14): The difference in magnitudes between blue (B) and yellow-green (V) light. It describes the color of a star. Blue stars have a negative color index, yellow stars have a B–V of 1–2.

Dec., Declination (p. 5): Angular distance from the celestial equator, positive towards the north. Equinox: 2000.0.

Dist., Distance (Fig. p. 2 and Table p. 3): Distances to stars and nebulae are listed in light years (ly) or million light years (M ly). A light year is the distance the light travels within one year, which is 9.46 trillion kilometers, 5.88 trillion miles, or 2100 times the distance Sun–Neptune. One parsec is 3.26 light years.

Eclipse (p. 16): The duration of the eclipse for eclipsing variable stars. It includes both decrease and increase of brightness. The time of minimum brightness is centered within this duration.

Extrema (p. 16): Maximum and minimum magnitudes of a variable star ever observed. These extreme values happen rarely.

Max., Maximum, Min., Minimum (p. 16): The time of maximum and minimum brightness of variable stars, given in the day numbering scheme of the Julian Date (see pp. 120, 121). "Min. = Max. + 10" means that the minimum occurs 10 days after the maximum.

2nd min. (p. 16): The magnitude at secondary minimum for variable stars.

Name (p. 9 and pp. 124, 125): A historic name of a star with international spelling preferred. Names still in use today are printed in bold letters.

Nebula (pp. 16–19 and 122, 123): A non-stellar object, designated by a number (NGC-number), by an IC-number or a Messier-number (M). Clusters of stars are included in the term "nebula" in this catalog. Descriptions for the 250 selected nebulae refer to 12×50 mm binoculars and a telescope of 150 mm (6 inch) aperture under a good, dark sky.

Period (p. 16): The duration of periodicity for variable stars, listed in days (d).

Position (p. 7): The location in the star chart, indicated in a small rectangle. A dots marks the location in the main stars charts. A small circle marks the location in the enlarged sections. Preceeding this rectangle is the abbreviated constellation (see pp. 126, 127). Dashed lines in the star charts show the boundaries between constellations.

PA, Position Angle (p. 15): The direction on the celestial sphere between both components of a binary star, graphically shown with north at the top.

R.A., Right Ascension (p. 5): Longitude on the celestial sphere, measured from the first point of aries or vernal point towards the east, given in hours and minutes from 0 to 24 hours ($1^h = 15°$). Equinox: 2000.0.

Sep., Separation (p. 15, Fig. p. 5, and Table p. 10): The apparent angular separation in arc-seconds between the two components of a binary star.

Shape (Table p. 19): Classification of nebulae according to their appearance in a telescope. This entry is preceded by an oval showing the elongation.

Size (Fig. p. 5 and Table p. 10): Apparent diameter of nebulae (along long axis), listed in arc-minutes.

Star (p. 9): Most stars are designated by a number according to Flamsteed and/or a Greek letter according to Bayer in combination with the genitive of the constellation. Data on brightness and color of stars include all components not resolved by the unaided eye, which are the components within five arc-minutes separation. The catalog contains 900 stars, including all 556 stars down to magnitude 4.0.

Star Charts (pp. 6, 7, and 132, 133): The main star charts contain all 5 700 stars down to magnitude 6.0 at a scale of $4°$/cm ($10°$/inch). The round magnified sections contain 30 000 out of the 200 000 stars down to magnitude 9.0 at a scale of $1°$/cm ($2.5°$/inch).

Te., Temperature (pp. 13, 14): Surface temperature of a star shown with a symbol of a thermometer. Hot stars are bluish: ↑, cool stars are yellow: ↓. Binaries with different temperatures display an impressive color contrast.

Type (pp. 16–19): Classification of nebulae into PN: planetary nebulae, DN: diffuse nebulae or galactic nebulae, OC: open star clusters, GC: globular star clusters, and Glx: galaxies.

Variable Star (pp. 15, 16): A star that changes its brightness. The catalog contains 80 variable stars with an amplitude of at least a quarter magnitude and a maximum brightness of at least about magnitude 5.

Vis., Visibility (Tables pp. 10 and 12, bottom): A measure of visibility of nebulae and binaries unique to this catalog. Six levels according to six instruments are distinguished. Objects of the visibility ⊡ are visible as a nebula or binary only in a telescope. Objects of the visibility ⊞ are visible as a nebula or binary even with the unaided eye. Open eyes in the symbols ⊡–⊞ indicate a low surface brightness which requires very dark sky for successful observation.

v-Mag., v-Magnitude (pp. 11–13, Fig. p. 4, and Tables p. 12): Visual magnitude according to the spectral sensitivity of the night-adapted eye with indirect vision. This catalog lists v-magnitudes for the total magnitude (first entry) and surface brightness (second entry, mag./□′) of nebulae. The surface brightness is the mean magnitude per square arc-minute.

V-Mag., V-Magnitude (p. 11, Fig. p. 4, and Table p. 12, top): The magnitude V in the UBV-system which closely corresponds to the spectral sensitivity of the eye with direct vision. This catalog lists V-magnitudes for stars. In most cases, the difference between v-magnitude and V-magnitude is insignificant. Variable stars have their typical magnitude range listed in the same column. A black dot preceeding the entry of the V-magnitude shows the size of the stellar disk in the main star chart.

Mean and Extreme Values

Parameter	Type	Median		Extreme	Object	Chart
Distance	nebula	9 000 ly	gr.	2 500 M ly	Quasar 3 C 273	E14
	galaxy	40 M ly	gr.	90 M ly	NGC 5746	E15
			sm.	180 000 ly	Large Magellanic Cl.	S3
	globular cluster	30 000 ly	gr.	80 000 ly	M 54, NGC 6715	E20
			sm.	7 000 ly	M 4, NGC 6121	E18
	open cluster	3 000 ly	gr.	12 000 ly	NGC 2324	E9
			sm.	150 ly	Hyades	E3
	diffuse nebula	3 000 ly	gr.	180 000 ly	Tarantula Nebula	S3
			sm.	1 200 ly	M 78, NGC 2068	E5
	planetary nebula	3 000 ly	gr.	6 000 ly	NGC 6818	E22
			sm.	500 ly	Helix Nebula	E24
	star	240 ly	gr.	6 000 ly	ϱ Cassiopeiae	N22
			sm.	4.40 ly	Toliman, α Centauri	S15
True Size	nebula	45 ly	sm.	0.1 ly	NGC 6572	E19
	galaxy	70 000 ly	gr.	250 000 ly	NGC 4565	E13
	globular cluster	60 ly	gr.	140 ly	ω Centauri	S15
	open cluster	18 ly	gr.	60 ly	h and χ Persei	N2
	diffuse nebula	25 ly	gr.	1 300 ly	Tarantula Nebula	S3
	planetary nebula	0.6 ly	gr.	3 ly	NGC 1360	E2
	binary-separation	2 light-days	gr.	5 ly	o Centauri	S12
			sm.	40 light-min.	Porrima, year 2006	E12
	star	7 million km	gr.	2 light-hours	μ Cephei	N22
			sm.	20 000 km	o^2 Eridani companion	E2
Apparent Size	nebula	$10'$	gr.	$420' = 7°$	Large Magellanic Cl.	S3
Brightness	nebula	mag. 8	gr.	mag. 0	Large Magellanic Cl.	S3
Luminosity	nebula	40 000 $\odot$	gr.	4 000 000 M $\odot$	Quasar 3 C 273	E14
	galaxy	22 000 M $\odot$	gr.	120 000 M $\odot$	M 49, NGC 4472	E14
	globular cluster	80 000 $\odot$	gr.	700 000 $\odot$	ω Centauri	S15
	open cluster	3 500 $\odot$	gr.	140 000 $\odot$	h and χ Persei	N2
	diffuse nebula	1 000 $\odot$	gr.	50 M $\odot$	Tarantula Nebula	S3
	planetary nebula	180 $\odot$	gr.	800 $\odot$	NGC 7662	N24
	star	$0.^{\mathrm{M}}2$, 70 $\odot$	gr.	100 000 $\odot$	Deneb, α Cyg $-7.^{\mathrm{M}}7$	N20
	$\odot$ = solar luminosity		sm.	0.000 4 $\odot$	Barnard's Star 13.^{M}2	E17
Color B–V		white 0.4 $\downarrow$	gr. red	3.4 •	R Leporis	E4
			sm. blue	-0.26 ⌡	ζ Puppis	S6
Color Contrast		0.4, 0.2 ↓↓	gr.	1.9, 0.0 •↓	Antares, α Scorpii	E18
Rotation	2000–2015	$1°$	gr.	$242°$	Porrima, γ Virginis	E12
Separation–Change	"	$0.''1$	gr.	$10.''0$	Toliman, α Centauri	S15
Proper Motion/Year		$0.''05$	gr.	$10.''4$	Barnard's Star	E17
Period	variable star	90 days	gr.	27 years	ε Aurigae	N6
			sm.	2 min.	Suhail Al Muhlif	S6
Amplitude	variable star	0.6	gr.	7 mag.	R Serpentis mag. 6–13	E16

Listed are mean value (median) and greatest/smallest value within the catalog.

Key to the Star Charts

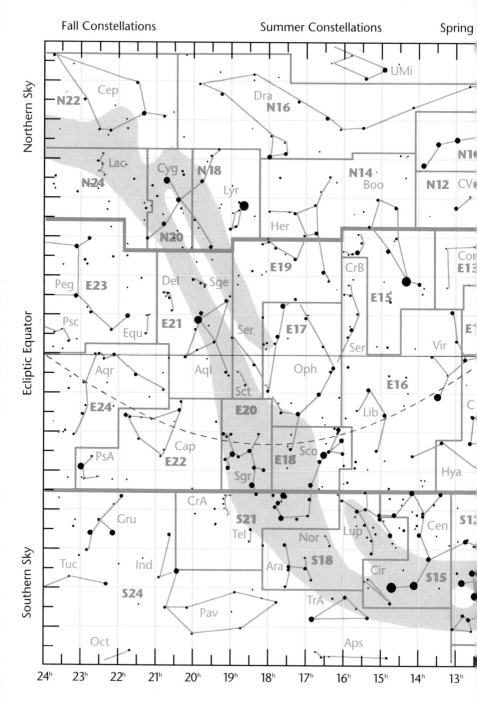

Fall Constellations   Summer Constellations   Spring

Northern Sky

Cep
N22
Lac
N24
Cyg
N18
Lyr
N20
Dra
N16
UMi
N14
Boo
N12
N10
N1
CV
Her
E19
CrB
E15
E13
Co
Peg
E23
Del
Sge
Ser
E17
E15
Psc
Equ
E21
Ser
Vir
E
Aqr
Aql
Oph
E24
Sct
E20
Lib
E16
PsA
Cap
E22
Sco
E18
Sgr
Hya
CrA
Gru
S21
Tel
Nor
Lup
Cen
S1
Ind
Ara
S18
Cir
S15
Tuc
S24
TrA
Pav
Oct
Aps

Southern Sky

Ecliptic Equator

24ʰ 23ʰ 22ʰ 21ʰ 20ʰ 19ʰ 18ʰ 17ʰ 16ʰ 15ʰ 14ʰ 13ʰ

Key to the Star Charts: This star chart indicates in green color the way
the celestial shere is divided into the 48 star charts on pp. 23-117.

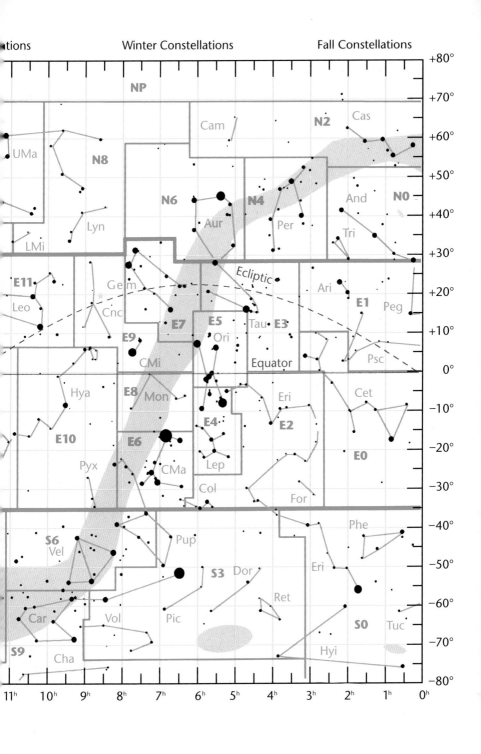